Study Guide and Student Solutions Manual for Tro's

Chemistry in Focus

A Molecular View of Our World

SECOND EDITION

Ann Tro

*University of California
at Santa Barbara*

BROOKS/COLE

™

THOMSON LEARNING

Australia • Canada • Mexico • Singapore • Spain • United Kingdom • United States

BROOKS/COLE

™

THOMSON LEARNING

Assistant Editor: *Stephanie Schmidt*
Marketing Assistant: *Stephanie Rogulski*
Editorial Assistant: *Faith Riley*
Production Coordinator: *Dorothy Bell*
Cover Design: *Terri Wright*

Cover Photo: *Darrell Gulin/CORBIS*
Cover Computer Graphics: Bob Western
Print Buyer: *Micky Lawler*
Printing and Binding: *Globus Printing*

For more information about this or any other Brooks/Cole product, contact:
BROOKS/COLE
511 Forest Lodge Road
Pacific Grove, CA 93950 USA
www.brookscole.com
1-800-423-0563 (Thomson Learning Academic Resource Center)

For permission to use material from this work, contact us by
Web: www.thomsonrights.com
fax: 1-800-730-2215
phone: 1-800-730-2214

Printed in the United States of America

10 9 8 7 6 5 4 3 2 1

ISBN 0-534-38030-1

To the Student

This solutions manual and study guide accompanies the textbook *Chemistry in Focus* by Nivaldo J. Tro. It will assist you in finding out whether you have learned the basic concepts. It organized by chapter with two parts to each chapter: Answers/Solutions and Review Tests.

The solutions manual portion of each chapter includes the answers to the odd-numbered questions and the solutions to the odd-numbered problems in the text.

The study guide portion of each chapter includes review tests and their answers. The review tests have matching questions, short answers questions, and multiple choice questions.

Best wishes for success in your study of chemistry.

CONTENTS

ANSWERS/SOLUTIONS TO

CHAPTER 1

MOLECULAR REASONS

ANSWERS TO QUESTIONS:

1. All natural phenomena in the world we can see are the result of molecular interactions we cannot see.

    Examples are numerous. Some suggestions are as follows:

    a) Ice melting to water.
    b) A match being struck produces a flame.
    c) A shirt fading when exposed to light or too many washings.

3. Chemistry is the science that investigates the molecular reason for the processes occurring in our macroscopic world.

5. A law is a concise statement or equation that summarizes a great variety of observations, while a theory explains the cause of the observations. A theory has been tested over a length of time and has more predictive power than a law.

7. The Greek approach to scientific knowledge was through pure reason and intuition. Today we approach science through experimentation.

9. The two main pursuits of alchemists were the transmutation of ordinary materials into gold and the discovery of the "elixir of life". Alchemists contributed to modern chemistry their understanding of metals, specifically how metals combine to form alloys. Alchemists also contributed laboratory separation and purification methods, including the isolation of natural substances (pharmacological) used to treat various ailments.

11. a) An element is a substance that cannot be broken down into simpler substances.

    b) A compound is a substance composed of two or more elements in a fixed definite proportion.

    c) A pure substance is matter that has a definite or constant composition and distinct properties.

d) A mixture is a combination of two or more substances in variable proportions.

13. Rutherford examined atomic structure through his gold foil experiment. By shooting alpha particles at a thin sheet of gold foil, Rutherford tested his idea that atoms were either soft, like blueberry muffins, or hard, like billiard balls. The results of his experiment were surprising. The majority of particles passed though the gold foil without deflection (as expected), but some deflected, scattering at angles both large and small. A small portion bounced back in the direction they had come from.

15. A black hole is a very dense form of matter where the structures of atoms have broken down, to form "solid" matter. The large mass and small size associated with the black hole causes a strong gravitational field, which allows mass and light to enter but not leave the black hole. Thus, because black holes neither reflect nor emit light, they are referred to as "black".

SOLUTIONS TO PROBLEMS:

17. a) Pure water would be classified a compound because it is composed of two or more elements in a fixed definite proportion.

   b) Copper wire is composed of the element copper.

   c) Graphite is composed of carbon atoms only and thus would be an element.

   d) Oil and water would be classified a heterogeneous mixture since a separation would occur into two distinct regions with different compositions.

19. The reaction of wood plus oxygen results in ash plus a variety of gases. As the reaction proceeds, the mass of the wood is decreasing, at the same time the mass of the products is increasing. The wood in a campfire eventually disappears but the mass is conserved because the products weigh exactly that of the wood plus oxygen.

21. Apply the law of conservation of mass:
          mass of reactants = mass of products

   a) 8.0 g natural gas + 32 g oxygen $\longrightarrow$ 17 g carbon dioxide + 16 g water

      40 g reactants $\neq$ 33 g products

   b) 5.7 g sodium + 8.9 g chlorine $\longrightarrow$ 14.6 g sodium chloride

      14.6 g reactants = 14.6 g products

23. The reaction must follow the law of conservation of mass

$$12 \text{ g hydrogen} + 104 \text{ g oxygen} \longrightarrow \_?\_ \text{ g water} + 8 \text{ g excess water}$$

$$116 \text{ g reactants} = 116 \text{ g products}$$
Therefore, the amount of water formed must be 108 g.

25. The law of constant composition says the ratio of hydrogen to carbon must always be the same.

a) $\dfrac{4.0 \text{ g hydrogen}}{12 \text{ g carbon}} = 0.33$

b) $\dfrac{1.5 \text{ g hydrogen}}{4.5 \text{ g carbon}} = 0.33$

c) $\dfrac{7.0 \text{ g hydrogen}}{17.0 \text{ g carbon}} = 0.41$ inconsistent, therefore incorrect

d) $\dfrac{10 \text{ g hydrogen}}{30 \text{ g carbon}} = 0.33$

27. In Rutherford's model of the atom, he proposed that the atom is electrically neutral. The number of positive charges (protons) would equal the number of negative charges (electrons).

     a) sulfur: 16 protons and 16 electrons
     b) bromine: 35 protons and 35 electrons

SOLUTIONS TO FEATURE PROBLEMS:

35. a) Salt is made up of the two elements sodium and chloride; it is a compound.

     b) Diamond is made up of only carbon; it is an element.

     c) Seawater consists of water, sodium ions, and chlorine ions; it is a homogeneous mixture.

     d) Snow is frozen water molecules; water consists of hydrogen and oxygen. Therefore, snow is a compound.

REVIEW TESTS FOR CHAPTER 1

Match the following statements or phrases to the end-of-chapter key terms.

1. A measurement of some aspect of nature. *observation*

2. In a chemical reaction, matter is neither created nor destroyed. *Law of Conservation of mass*

3. A tentative model that describes the underlying cause of physical behavior. *theory*

4. The process by which a set of observations leads to a model of reality. *Scientific Method*

5. The Greek philosopher who stated "Water is the principle, or the element of things. All things are water." *Empedocles*

6. The astronomer who first claimed the sun to be the center of the universe. *Copernicus*

7. The scientist who proposed that if a substance could be broken down into simpler substances, it was not an element. *Boyle*

8. A substance composed of two or more elements in fixed definite proportions. *Compound*

9. A specific type of mixture which can be separated into two or more regions with different compositions. *heterogeneous mixture*

10. The French chemist who established the law of conversation of mass. *Lavoisier*

11. The law that states all samples of a given compound have the same proportions of their constituent elements. *Law of Constant Composition*

12. The British scientist who formulated the Atomic Theory. *Dalton*

13. The internal structure of the atom was studied by this scientist. *Rutherford*

14. Positively charged particles present in the nucleus of the atom. *protons*

ANSWERS TO MATCHING

1. observation
2. Law of conservation of mass
3. theory
4. scientific method
5. Empedocles
6. Copernicus
7. Boyle
8. compound
9. heterogeneous mixture
10. Lavoisier
11. law of constant composition
12. Dalton
13. Rutherford
14. protons

SELF-TEST QUESTIONS

Completion: Write the word, phrase, or number that will complete the statement or answer the question in the blank.

1. All natural phenomena that occurs in the world around us can be traced backs to a _____ cause.

2. A theory is tested by _____.

3. Empedocles suggested that all matter was composed of four basic materials or elements: Air, water, fire and _____.

4. In Dalton's Atomic Theory, each element is composed of particles called _____.

5. Rutherford examined atomic structure by directing small particles called _____, at a thin sheet of gold foil.

6. The number of protons in the nucleus of a neutral atom is always equal to the number of _____ outside the nucleus.

Multiple Choice: Select the correct answer from the choices listed.

7.   What is the term given to series of related measurements that are then combined to formulate a broadly applicable generalization?

     a)   observation
     b)   experiment
     c)   theory
     d)   scientific law

8.   What is the name of the anatomist who was able to portray accurate observations of human anatomy?

     a)   Boyle
     b)   Vesalius
     c)   Aristotle
     d)   Proust

9.   Which one of the following substances would be classified as an element?

     a)   water
     b)   sulfur
     c)   sugar
     d)   air

10.   All of the following would be examples of pure substances except

     a)   air
     b)   hydrogen
     c)   salt
     d)   carbon dioxide

11.   A combination of sand, salt, and water is an example of a

     a)   homogeneous mixture
     b)   heterogeneous mixture
     c)   compound
     d)   pure substance

12. Several samples of nitrogen dioxide are obtained and decomposed into nitrogen and oxygen. Which one of the results does not follow the law of constant composition?

   a)   28 g of nitrogen and 32 g of oxygen
   b)   3.5 g of nitrogen and 4.0 g of oxygen
   c)   7.0 g of nitrogen and 9.0 g oxygen
   d)   39.2 g of nitrogen and 44.8 g of oxygen

13. Which one of the following statements is part of Dalton's Atomic Theory?

   a)   The atoms contained in elements are in constant motion.
   b)   Atoms of a given element can have different masses.
   c)   In nucleus reactions, atoms can be created.
   d)   Atoms of different elements can combine to form compounds.

14. Which one of the following statements is not consistent with Rutherford's theory of the nucleus atom?

   a)   Most of the volume of the atom is empty space occupied by negatively charged electrons.
   b)   Most of the mass of the atom is contained in a small space called the nucleus.
   c)   The electrons travel in specific orbits around the nucleus of the atom.
   d)   Protons impart the positive charge to the nucleus of the atom.

15. Which statement best characterizes black holes in the universe?

   a)   Black holes can both reflect and emit light.
   b)   Black holes repel all things in their general vicinity.
   c)   Black holes cause very weak gravitational fields.
   d)   Black holes have large masses and small sizes.

ANSWERS TO SELF-TEST QUESTIONS

1. molecular or atomic
2. experiment
3. earth
4. atoms
5. alpha
6. electrons
7. d
8. b
9. b
10. a
11. b
12. c
13. d
14. c
15. d

CHAPTER 2

THE CHEMIST'S TOOL BOX

ANSWERS TO QUESTIONS:

1. Curiosity is an important part of the scientific enterprise because scientists need a strong desire to investigate and learn about the behavior of nature. Science must start with the question why. The scientific method is then utilized to accumulate systematized knowledge about the physical world. A scientist's curiosity is incapable of being satisfied. Without this curious nature of scientist, the advancement of science would not have occurred as we presently know it.

3. Theorizing or model building allows scientists to continue asking why and perform more experimentation concerning a theory or model. Through the model-building process, new information is obtained. From the new information, some questions are answered, and new questions may be raised.

5. Units are important because they give meaning to the measured quantity. Without units, a measurement would represent an abstract number and give no useful information.

7. Answers may vary. Three possible units for length are grams, milligrams and kilograms.
    Examples:
    Mass of a penny - grams (g)
    Mass of a straight pin - milligrams (mg)
    Mass of a bucket of water - kilograms (kg)

9. Answers may vary. Three possible units for volume are milliliters, kiloliters (gallons), and liters.
    Examples:
    Volume of a child's juice box – milliliters (mL)
    Volume of water in a swimming pool – gallons or kiloliters (kL)
    Volume of a bottle of soda – liters (L)

11. A theory or model is an interpretation – speculation – to explain why nature behaves in a particular way. The interpretations are only models of reality and are not exact. With more knowledge theories can change and grow to fit the new data.

13. For the number 3.4×10^6
    Decimal – 3.4
    Exponential – 10^6
    Exponent – 6

15. a) The total increase in carbon monoxide:

$$330\,\text{ppm} - 280\,\text{ppm} = 50\,\text{ppm}$$

b) The average yearly increase in carbon monoxide:

$$\frac{50\,\text{ppm}}{(2000-1800)\text{yrs}} = .25\,\text{ppm}/\text{yr}$$

c) The total percentage increase over this period:

$$\left(\frac{50\,\text{ppm}}{280\,\text{ppm}}\right) \times 100\% = 18\%$$

d) The average yearly percentage increase:

$$\frac{18\%}{200\text{yrs}} = .09\%/\text{yr}\ (.09\%\,\text{per year})$$

SOLUTIONS TO PROBLEMS:

17. a) 3.2667×10^7 people
    b) 5.926467×10^9 people
    c) 7.461×10^{-11} m
    d) 1.5×10^{-5} m

19. a) 602,200,000,000,000,000,000,000 carbon atoms / 12.01 grams of carbon
    b) 299,000,000 m/s
    c) 0.000000450 m
    d) 16,000,000,000 years

21. a) 9.2×10^8
    b) 8.7×10^{-2}
    c) 6.2×10^4
    d) 2.1×10^{-6}

23. a) 1.6×10^8
    b) 3.6×10^{35}
    c) 2.5×10^{-6}
    d) 1.2×10^7

25. a) $x \cong 6.2$ (two significant figures)
    b) $x \cong 0.83$ (two significant figures)
    c) $x \cong -0.268$ (three significant figures)
    d) $x \cong 0.28$ (two significant figures)
    e) $x \cong \pm 8.57$ (three significant figures)

27. a) $8.95 \, \text{cm} \times \dfrac{1 \, \text{m}}{100 \, \text{cm}} = 0.0895 \, \text{m}$

    b) $1298.4 \, \text{g} \times \dfrac{1 \, \text{kg}}{1000 \, \text{g}} = 1.2984 \, \text{kg}$

    c) $129 \, \text{cm} \times \dfrac{10 \, \text{mm}}{1 \, \text{cm}} = 1290 \, \text{mm}$

    d) $2452 \, \text{mL} \times \dfrac{1 \, \text{L}}{1000 \, \text{mL}} = 2.452 \, \text{L}$

29. a) $40 \, \text{cm} \times \dfrac{1 \, \text{in.}}{2.54 \, \text{cm}} = 16 \, \text{in.}$

    b) $27.8 \, \text{m} \times \dfrac{3.28 \, \text{ft}}{1 \, \text{m}} = 91.2 \, \text{ft}$

    c) $10 \, \text{km} \times \dfrac{0.62 \, \text{mi}}{1 \, \text{km}} = 6.2 \, \text{mi}$

    d) $3800 \, \text{m} \times \dfrac{3.28 \, \text{ft}}{1 \, \text{m}} = 12{,}500 \, \text{ft}$

31. a) $1 \, \text{m}^2 \times \left(\dfrac{100 \, \text{cm}}{1 \, \text{m}} \right)^2 = 1 \times 10^4 \, \text{cm}^2$

    b) $1 \, \text{yd}^3 \times \left(\dfrac{3 \, \text{ft}}{1 \, \text{yd}} \right)^3 \times \left(\dfrac{12 \, \text{in.}}{1 \, \text{ft}} \right)^3 = 4.7 \times 10^4 \, \text{in.}^3$

    c) $1 \, \text{ft}^2 \times \left(\dfrac{12 \, \text{in}}{1 \, \text{ft}} \right)^2 \times \left(\dfrac{2.54 \, \text{cm}}{1 \, \text{in.}} \right)^2 = 9.3 \times 10^2 \, \text{cm}^2$

33. $d = \dfrac{m}{V} = \dfrac{51.4 \, \text{g}}{3.80 \, \text{mL}} = 13.5 \, \text{g} / \text{mL}$

35. a) **Mass of Gold**

$$350\,mL \times \frac{1\,cm^3}{1\,mL} \times \frac{19.32\,g}{1\,cm^3} = 6.8 \times 10^3 \text{ g of gold}$$

Mass of Sand

$$350\,mL \times \frac{1\,cm^3}{1\,mL} \times \frac{3.00\,g}{1\,cm^3} = 1.0 \times 10^3 \text{ g of sand}$$

b) Yes, the woman would notice the change from gold to sand since the sand weighs much less than the same volume of gold.

37. a) $m = 1.7 \times 10^{-24}\,g \qquad r = 1 \times 10^{-13}\,cm$

$$V = \frac{4}{3}\pi r^3 = \frac{4}{3} \times 3.14 \times \left(1 \times 10^{-13}\right)^3 = 4 \times 10^{-39}\,cm^3$$

$$d = \frac{m}{V} = \frac{1.7 \times 10^{-24}\,g}{4 \times 10^{-39}\,cm^3} = 4 \times 10^{14}\,g/cm^3$$

b) $r = 1 \times 10^{-4}\,m \times \dfrac{100\,cm}{1\,m} = 1 \times 10^{-2}\,cm$

$$V = \frac{4}{3}\pi r^3 = \frac{4}{3} \times 3.14 \times \left(1 \times 10^{-2}\,cm\right)^3 = 4 \times 10^{-6}\,cm^3$$

Mass of the black hole :

density from part a : $d = 4 \times 10^{14}\,g/cm^3$

$$4 \times 10^{-6}\,cm^3 \times \frac{4 \times 10^{14}\,g}{1\,cm^3} = 1.6 \times 10^9\,g \times \frac{1\,kg}{1000\,g} = 1.6 \times 10^6\,kg$$

SOLUTIONS TO FEATURE PROBLEMS:

43. The most accurate scale has the divisions every 1/10 lb. The least accurate scale has the divisions every 10 lb. To distinguish the accuracy of the scales you could write the weight for Figure E to the second decimal place, the weight for Figure F to the first decimal place, and for Figure G you would have to write the answer with no decimal. For example the weights from Figures E-G might be 2.55 lb, 2.5 lb, and 2 lb respectively. In all cases the uncertainty of the measurement is in the last digit.

REVIEW TESTS FOR CHAPTER 2

Match the following statements or phrases to the end-of-chapter key terms.

1. The SI standard unit of length.

2. The SI standard unit of mass.

3. Any two quantities known to be equal.

4. The mass-to-volume ratio of an object.

5. The SI standard unit of time.

ANSWERS TO MATCHING

1. meter
2. kilogram
3. conversion factor
4. density
5. second

SELF-TEST QUESTIONS

Completion: Write the word, phrase, or number in the blank that will complete the statement or answer the question.

1. Identify the exponent in the number, 5.46×10^6.

2. How many significant figures in 132.06?

3. The prefix "milli" in the metric system means _____ 10^{-3} or .001

4. The symbol "μ" is assigned what prefix name? micro

5. Three liters is the same as _____ mL.

6. Convert 32.6 m to ft.

    $3.28 \text{ ft} = \text{m}$ $32.6 \text{ m} \times \dfrac{3.28}{\text{m}}$

    $= 107 \text{ ft}$

7. Convert 6.52×10^{-3} to decimal notation.

8. Perform each of the following operations.

    a. $43 \times 10^{-3} + 26 \times 10^{-2}$
    b. $85 \times 10^{3} \times 6.4 \times 10^{2}$
    c. $4 \times 10^{6}/4x \ 10^{-6}$

9. Solve the following equation for x:

$4x + 10 = 19$

10. Convert 15 kg to Mg.

11. A particular fad diet claims a weight loss of 3 pounds/week. How many grams/day is this?

12. A 25.2 g sample of magnesium has a volume of 14.5 cm^3. What is the density of magnesium in g/cm^3?

Multiple Choice: Select the correct answer from the choices listed.

13. In scientific notation, 0.0753 =

    a) 75.3×10^{3}
    b) 75.3×10^{-3}
    c) 7.53×10^{-2}
    d) 7.53×10^{2}

14. Which one of the following units does not represent a volume measurement?

    a) m^3
    b) cm^2
    c)   mL
    d)   L

15. Which one of the following measurements contains 2 significant figures?

    a)   0.02 g
    b)   0.020 g
    c)   12.0 g
    d)   0.2 g

16. The metric system is based on multiples of:

a) 0.1
b) 1
c) 10
d) 100

17. Which of the following expressions is not a conversion factor?

a) 35 cm^3
b) 1 L = 1.057 quarts
c) 1 m = 39.37 in
d) 55 mi/hr

18. Given the following densities (in g/cm^3), select the one that floats on water.

a) 0.57
b) 1.21
c) 2.70
d) 10.18

19. 10^{-2} can also be written as:

a) 100
b) 0.001
c) 1/20
d) 1/100

20. The density of copper is 8.96 g/mL. What mass (in g) of copper occupies 10.0 mL?

a) 1.12
b) 89.6
c) 0.896
d) 896

ANSWERS TO SELF-TEST QUESTIONS

1.   6
2.   5
3.   0.001 or 10^{-3}
4.   micro
5.   3000
6.   107 ft
7.   0.00652
8.   a. 0.30 b. 5.4×10^7 c. 1×10^{12}
9. $x = 9/4$
10.   0.015 Mg
11.   194 g/day
12.   174 g/cm^3
13.   c
14.   b
15.   b
16.   c
17.   a
18.   a
19.   d
20.   b

CHAPTER 3

ATOMS AND ELEMENTS

ANSWERS TO QUESTIONS:

1. Atoms are the building blocks of matter. If we are to understand the macroscopic properties of an element, we must first understand the microscopic properties of its atoms. Properties studied on an atomic scale correlate directly with properties of the element on a macroscopic scale.

3. There are 91 naturally occurring elements. The number of protons in the nucleus defines the element.

5.

Chemical Symbol	Name	Atomic Number
H	Hydrogen	1
He	Helium	2
Li	Lithium	3
Be	Beryllium	4
B	Boron	5
C	Carbon	6
N	Nitrogen	7
O	Oxygen	8
F	Fluorine	9
Ne	Neon	10
Na	Sodium	11
Mg	Magnesium	12
Al	Aluminum	13
Si	Silicon	14
P	Phosphorus	15
S	Sulfur	16
Cl	Chlorine	17
Ar	Argon	18
Fe	Iron	26
Cu	Copper	29
Br	Bromine	35
Kr	Krypton	36
Ag	Silver	47
I	Iodine	53
Xe	Xenon	54
W	Tungsten	74
Au	Gold	79
Hg	Mercury	80
Pb	Lead	82

Chemical Symbol	Name	Atomic Number
Rn	Radon	86
U	Uranium	92

7.

	Mass (g)	Mass (amu)	Charge
Proton	1.67×10^{-24}	1	+1
Neutron	1.67×10^{-24}	1	0
Electron	9.11×10^{-28}	0	-1

9. Mendeleev's biggest contribution to modern chemistry was his placement of the elements on the periodic table and thus the periodic law. He realized that arranging the elements in order by their atomic number "...certain sets of properties reoccur periodically." From this he predicted the chemical and physical properties of undiscovered elements and unknown compounds.

11. There are numerous examples for the following terms. Some suggestions are as follows:

Alkali metal	Li, K
Alkaline earth metal	Ca, Sr
Halogen	F, I
Noble gas	He, Rn
Metal	Mg, Ag
Non-metal	C, P
Transition Metal	Fe, Au
Metalloid	B, Sb

SOLUTIONS TO PROBLEMS:

13.

		Protons	Electrons
a)	Na^+	11	10
b)	O^{2-}	8	10
c)	Cr^{2+}	24	22
d)	I^-	53	54
e)	Fe^{3+}	26	23

15.

a)	oxygen:	$Z = 8$	$A = 16$
b)	chlorine:	$Z = 17$	$A = 35$
c)	sodium:	$Z = 11$	$A = 23$
d)	uranium:	$Z = 92$	$A = 235$

17. a) $^{235}_{92}U$

   b) $^{238}_{92}U$

   c) $^{239}_{94}Pu$

   d) $^{144}_{54}Xe$

19. a) protons = 94
       neutrons = 145
       electrons = 94

   b) protons = 24
       neutrons = 28
       electrons = 21

   c) protons = 8
       neutrons = 8
       electrons = 10

   d) protons = 20
       neutrons = 20
       electrons = 18

21. a) Helium $n = 1$ $2e^-$

   b) Aluminum $n = 1$ $2e^-$
                      $n = 2$ $8e^-$
                      $n = 3$ $3e^-$

   c) Beryllium $n = 1$ $2e^-$
                      $n = 2$ $2e^-$

   d) Neon $n = 1$ $2e^-$
                      $n = 2$ $8e^-$

   e) Oxygen $n = 1$ $2e^-$
                      $n = 2$ $6e^-$

23. a) He: 2 valence e^-
    b) Al: 3 valence e^-
    c) Be: 2 valence e^-
    d) Ne: 8 valence e^-
    e) O: 6 valence e^-

25.

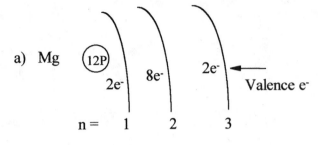

a) Mg

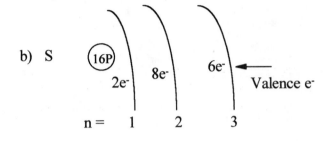

b) S

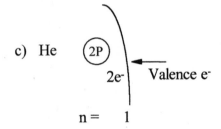

c) He   (2P)

2e⁻   ←   Valence e⁻

n =   1

27. Na and Li because they are both alkali metals.
Ne and Ar because they are both noble gases.
O and S because they are both in group 6.

29. Mg^{2+} $n = 1$ 2e⁻
           $n = 2$ 8e⁻
           $n = 3$ 0e⁻

The magnesium ion (Mg^{2+}) is more stable, less reactive, than the neutral Mg because the ion has lost 2 electrons to have an octet in its $n = 2$ shell, which is now its outer shell. The reactivities of F and F⁻ can be compared similarly. The neutral F is more reactive because it would like to gain an electron in order to have an octet in its outer shell.

31. a) Mn – metal
    b) I – nonmetal
    c) Te – metalloid
    d) Sb – metalloid
    e) O – nonmetal

33. Given: Isotope 1 – mass = 106.905 amu
   – relative abundance (fraction of isotope) = 51.8% (or 0.518)

   Isotope 2 – mass = 108.904 amu
   – relative abundance (fraction of isotope) = 48.2% (or 0.482)

Atomic mass = (fraction of isotope 1 × mass of isotope 1) + (fraction of isotope 2 × mass of isotope 2)
Atomic mass = (0.518 × 106.905) + (0.482 × 108.904) = 107.9

   The element is silver (Ag).

35. Given: Isotope 1 – mass = 62.939 amu
   – relative abundance (fraction of isotope) = 69.17% (or 0.6917)

Relative abundance of Isotope 2 = 100% − 69.17% = 30.83% (the percents of all natural isotopes must add up to 100%)

Atomic mass = (fraction of isotope 1 × mass of isotope 1) + (fraction of isotope 2 × mass of isotope 2)

   Let x = mass of isotope 2

   Substitute numbers into equation,

   63.55 amu = (0.6917 × 62.939) + (.3083x)

   Solve for x,

   63.55 - 43.53 = .3083x

   $$x = \frac{20.02 \text{ amu}}{.3083}$$

   x = 64.94 amu = mass of isotope 2

37. $$2.6 \text{ kg Al} \times \frac{1000 \text{ g Al}}{1 \text{ kg}} \times \frac{1 \text{ mol Al}}{26.98 \text{ g}} = 96 \text{ moles Al}$$

39. a) $$7.30 \text{ mol P} \times \frac{30.97 \text{ g P}}{1 \text{ mol P}} = 226 \text{ g P}$$

   b) $$14.8 \text{ mol C} \times \frac{12.01 \text{ g C}}{1 \text{ mol C}} = 178 \text{ g C}$$

   c) $$5.00 \times 10^{23} \text{ Cr atoms} \times \frac{1 \text{ mol Cr}}{6.02 \times 10^{23} \text{ Cr atoms}} \times \frac{52.00 \text{ g Cr}}{1 \text{ mol Cr}} = 43.2 \text{ g Cr}$$

   d) $$2.80 \times 10^{24} \text{ I atoms} \times \frac{1 \text{ mol I}}{6.02 \times 10^{23} \text{ I atoms}} \times \frac{126.9 \text{ g I}}{1 \text{ mol I}} = 590 \text{ g I}$$

41. $4.8 \text{ g Pt} \times \dfrac{1 \text{ mol Pt}}{195.1 \text{ g Pt}} \times \dfrac{6.02 \times 10^{23} \text{ atoms Pt}}{1 \text{ mol Pt}} = 1.5 \times 10^{22} \text{ atoms Pt}$

43. $32 \text{ cm}^3 \text{ Ti} \times \dfrac{4.50 \text{ g Ti}}{1 \text{ cm}^3 \text{ Ti}} \times \dfrac{1 \text{ mol Ti}}{47.88 \text{ g Ti}} \times \dfrac{6.02 \times 10^{23} \text{ atoms Ti}}{1 \text{ mol Ti}} = 1.8 \times 10^{24} \text{ atom Ti}$

45. Given : $r_{sun} = 7 \times 10^8 \text{ m} \times \dfrac{100 \text{ cm}}{1 \text{ m}} = 7 \times 10^{10} \text{ cm}$

$d_{sun} = 1.4 \text{ g/cm}^3$

$V = \dfrac{4}{3} \pi r^3$

a) $V_{sun} = \dfrac{4}{3}(3.14)\left(7 \times 10^{10} \text{ cm}\right)^3 = 1 \times 10^{33} \text{ cm}^3$

$m_{sun} = 1 \times 10^{33} \text{ cm}^3 \times \dfrac{1.4 \text{ g}}{1 \text{ cm}^3} = 1 \times 10^{33} \text{ g}$

Since the sun consists mostly of Hydrogen we can calculate the number of atoms in the sun.

$1 \times 10^{33} \text{ g H} \times \dfrac{1 \text{ mol H}}{1.00 \text{ g H}} \times \dfrac{6.02 \times 10^{23} \text{ atoms H}}{1 \text{ mol H}} = 6 \times 10^{56} \text{ atoms H in the sun}$

b) 6×10^{56} represents the number of atoms in a star.

$\dfrac{6 \times 10^{56} \text{ atoms}}{1 \text{ star}} \times \dfrac{1 \times 10^{11} \text{ stars}}{1 \text{ galaxy}} = 6 \times 10^{67} \text{ atoms/galaxy}$

$\dfrac{6 \times 10^{67} \text{ atoms}}{1 \text{ galaxy}} \times \dfrac{1 \times 10^{9} \text{ galaxy}}{1 \text{ universe}} = 6 \times 10^{76} \text{ atoms in the universe}$

c) This number, 1×10^{23}, is quite small compared with the number of atoms in the universe, 6×10^{76}.

SOLUTIONS TO FEATURE PROBLEMS:

53. Atoms (a) have an atomic weight less than (b) while atoms (c) have an atomic weight more than (b).

$$175 \text{ g (b)} \times \frac{1 \text{ dozen}}{25 \text{ g (b)}} \times \frac{12 \text{ atoms}}{1 \text{ dozen}} = 84 \text{ atoms of (b)}$$

REVIEW TESTS FOR CHAPTER 3

Match the following statements or phrases to the end-of-chapter key terms.

1. The smallest identifiable unit of an element.

2. The number of protons located in the nucleus of an atom.

3. Positively charged ions.

4. The term used to describe atoms having the same number of protons but different numbers of neutrons.

5. The sum of the number of neutrons and protons in the nucleus of a given atom.

6. The assigned average mass to an element.

7. The theory that explains why the properties of elements recur in a periodic fashion.

8. The electrons in the outer orbit of an element's atoms.

9. The family name of the elements in group 8A on the periodic table.

10. The family name of the elements in the second column of the periodic table.

11. The name of the group 7A elements on the periodic table; these elements are only one electron short of a stable electron configuration.

12. The specific name given to the metals on the periodic table that lose electrons in chemical reactions, but do not necessarily acquire a noble gas configuration.

13. The concept that relates the mass and the number of atoms in a sample.

14. The name given to the convenient number that chemists use in dealing with atoms.

15. The number that specifies each orbit in the electron configuration of an atom.

ANSWERS TO MATCHING

1. atom
2. atomic number
3. cations
4. isotopes
5. mass number
6. atomic weight
7. Bohr Model
8. valance electrons
9. noble gases
10. alkaline-earth metals
11. halogens
12. transition metals
13. mole concept
14. Avogadro's number
15. quantum number

SELF-TEST QUESTIONS

Completion: Write the word, phrase, or number in the blank that will complete the statement or answer the question.

1. What is the name and the atomic number that corresponds to the chemical symbol Rn?

2. Write down the chemical symbol for the element tungsten.

3. For the isotope uranium-238, write its symbol in the form $_Z^A X$.

4. How many electrons are in the sulfide ion, S^{2-}?

5. Draw the electron configuration for the element phosphorus according to the Bohr model.

6. The element nitrogen has _____ valence electrons.

Multiple Choice: Select the correct answer from the choices listed.

7. The element mercury (Hg) is classified as a(n):

   a) alkali metal
   b) halogen
   c) transition metal
   d) metalloid

8. Use the following information to identify the atom or ion:

Mass number	Protons	Neutrons	Electrons
28	14	14	10

   a) Si^{4+}
   b) $Si4^{4-}$
   c) Si
   d) Ni^{4+}

9. Consider the following elements. Which one would you expect to be the most reactive?

   a) K
   b) Ca
   c) Zn
   d) Kr

10. In the Bohr model of the atom, the higher the quantum number, the _____ the distance between the electron and the nucleus and the _____ its energy.

    a) lower, lower
    b) higher, higher
    c) lower, higher
    d) higher, lower

11. The subatomic particle with a charge of -1 and a mass of 9×10^{-28} g is:

    a) nucleus
    b) proton
    c) neutron
    d) electron

12. Which one of the following elements has a full outer orbit in its electron configuration?

    a) bromine
    b) helium
    c) hydrogen
    d) oxygen

13. Which one of these elements would you expect to be most chemically similar to calcium?

    a) magnesium
    b) potassium
    c) scandium
    d) sodium

14. One mole of sulfur contains how many sulfur atoms?

    a) 1
    b) 32.06
    c) 6.022×10^{23}
    d) 6.022×32.06

15. In Mendeleev's periodic table, elements are arranged in order of increasing:

    a) atomic number
    b) atomic weight
    c) number or protons
    d) number of isotopes

16. The number of _____ in the nucleus of an atom determines what the element is.

    a) neutrons
    b) electrons
    c) protons
    d) protons and electrons

17. Which one of the following chemical symbols is not correctly written?

    a) Ne
    b) Cl
    c) Be
    d) PB

18. The calcium ion (Ca^{2+}) has an electron configuration identical to:

a) Kr
b) Ar
c) Ca
d) Ti

19. The total electrical charge on the nucleus is determined by what specific information of the atom?

a) atomic weight
b) number of neutrons
c) atomic number
d) mass number

20. Which kind of elements primarily gain electrons in chemical reactions?

a) metals
b) non-metals
c) metalloids

ANSWERS TO SELF-TEST QUESTIONS

1. radon, 86
2. W
3. $^{238}_{92}U$
4. 16
5. n=1 2e; n=2 8e; n=3 5e
6. 5
7. c
8. a
9. a
10. b
11. d
12. b
13. a
14. c
15. b
16. c
17. d
18 b
19. c
20. b

CHAPTER 4

MOLECULES AND COMPOUNDS

ANSWERS TO QUESTIONS:

1. Atoms are rarely found in nature in an uncombined state. Because most atoms do not have complete octets in their outer shell they are very reactive. The result of their reactivity is the formation of compounds. All matter is made of atoms; the atoms are found within compounds and compounds combine to form mixtures. Therefore it is consistent to say that all matter is made of atoms and that most common substances are either compounds or mixtures.

3. A chemical formula represents a compound or a molecule. The symbols for the elements are used to indicate the types of atoms present and subscripts are used to indicate the relative number of atoms.

5. The transfer of valence electrons between a metal and a nonmetal forms ionic bonds. The bonding is a result of electrostatic attraction between a positively charged ion (metal) and a negatively charged ion (nonmetal). A crystalline structure of positive and negative ions results. The sharing of valence electrons between two nonmetals forms covalent bonds. Covalent compounds are composed of clusters of two or more atoms bonded together to form molecules.

7. The properties of compounds are determined by the shape and structure of the molecule, the kinds of atoms present, and the types of bonds present.

9. A chemical formula reveals equivalencies (conversion factors) between elements in a particular compound. These equivalencies enable us to calculate the amounts of the constituent elements present in a given amount of the compound. For example, in 1 mole of CO_2, there is 1 mole of carbon and 2 moles of oxygen.

11. In a chemical reaction, the reactants are the substances on the left-hand side of the equation – the substances you start with. The products are the substances on the right-hand side of the equation – the substances you end up with.

13. The relative numbers of reactants and products in a reaction are given by the coefficients in the balanced chemical equation. These coefficients represent conversion factors similar to the subscripts in chemical formulas. For example in the equation: $CH_4 + 2O_2 \longrightarrow CO_2 + 2H_2O$ the coefficients tell us that for every 1 mole of CH_4 and 2 moles of O_2 that combine, 1 mole of CO_2 and 2 moles of H_2O will be produced. These conversion factors can be used to predict the amount of reactant(s) needed in a particular reaction, or the amount of a particular product(s) that will form.

SOLUTIONS TO PROBLEMS:

15. The molecular weight is calculated by adding the atomic weights of all the atoms in the compound. Remember, if there is a subscript other than 1, that number must be multiplied by the atomic weight of that particular atom and then added to the total.

    a) NaCl   58.44 amu or 58.44 g/mol
    b) CH_4 16.05 amu or 16.05 g/mol
    c) NH_4Cl 53.50 amu or 53.50 g/mol
    d) $C_{12}H_{22}O_{11}$ 342.34 amu or 342.34 g/mol

17. To determine the molecular formula, first divide the total molecular weight by the atomic weight of the heavier atom.

$$\frac{70.1\,\text{amu}}{12.01\,\text{amu of C}} = 5.84$$

We see that the number is not even - it shouldn't be since we have to account for the hydrogens. But now we know that their are 5 carbons. The next step is to figure out the molecular weight of 5 carbons.

$$5 \times 12.01 = 60.05\,\text{amu}$$

Now we need to determine the number of hydrogens.

$$70.1 - 60.05 = 10.05\,\text{amu (the weight of the total number of hydrogens)}$$

$$\frac{10.05}{1.008\,\text{(the atomic weight of one H)}} = 10\,\text{hydrogens}$$

Therefore, the molecular formula is C_5H_{10}.

19. $0.5\,\text{g}\,C_3H_6O \times \dfrac{1\,\text{mol}\,C_3H_6O}{58.09\,\text{g}\,C_3H_6O} \times \dfrac{6.02 \times 10^{23}\,\text{molecules}}{1\,\text{mol}} = 5 \times 10^{21}\,\text{acetone molecules}$

21.

    a) $0.050\,\text{cm}^3 \times \dfrac{1\,\text{gram}}{1\,\text{cm}^3} \times \dfrac{1\,\text{mole}}{18.02\,\text{g}} \times \dfrac{6.02 \times 10^{23}\,\text{molecules}}{1\,\text{mole}} = 1.7 \times 10^{21}\,\text{water molecules}$

    b) $1.7 \times 10^{21}\,H_2O\,\text{molecules} \times \dfrac{2\,\text{atoms H}}{1\,\text{molecule}\,H_2O} = 3.4 \times 10^{21}\,\text{H atoms}$

23.

a) $5\,H_2O \text{ molecules} \times \dfrac{2\,H \text{ atoms}}{1\,H_2O \text{ molecule}} = 10\,H \text{ atoms}$

b) $58\,CH_4 \text{ molecules} \times \dfrac{4\,H \text{ atoms}}{1\,CH_4 \text{ molecule}} = 232\,H \text{ atoms}$

c) $1.4\times10^{22}\,C_{12}H_{22}O_{11} \text{ molecules} \times \dfrac{22\,H \text{ atoms}}{1\,C_{12}H_{22}O_{11} \text{ molecule}} = 3.1\times10^{23}\,H \text{ atoms}$

d) $14 \text{ dozen }NH_3 \text{ molecules} \times \dfrac{3\,H \text{ atoms}}{1\,NH_3 \text{ molecule}} \times \dfrac{12 \text{ molecules}}{1 \text{ dozen}} = 504\,H \text{ atoms}$

25.

a) $0.58 \text{ mol }N_2 \times \dfrac{2 \text{ mol }N}{1 \text{ mol }N_2} = 1.2 \text{ mol }N$

b) $1.7 \text{ mol }NO_2 \times \dfrac{1 \text{ mol }N}{1 \text{ mol }NO_2} = 1.7 \text{ mol }N$

c) $0.12 \text{ mol }N_2O \times \dfrac{2 \text{ mol }N}{1 \text{ mol }N_2O} = 0.24 \text{ mol }N$

d) $1.3 \text{ mol }N_2H_4 \times \dfrac{2 \text{ mol }N}{1 \text{ mol }N_2H_4} = 2.6 \text{ mol }N$

27. a) strontium oxide
 b) beryllium chloride
 c) magnesium sulfate
 d) calcium chloride
 e) sodium hydroxide

29. a) sulfur hexafluoride
 b) dinitrogen tetraoxide
 c) sulfur dioxide

31. a) SO_3
 b) PCl_5
 c) CS_2
 d) $SiCl_4$
 e) KOH

33. a) $3 PbO + 2 NH_3 \longrightarrow 3 Pb + N_2 + 3 H_2O$

b) $Mg_3N_2 + 6 H_2O \longrightarrow 3 Mg(OH)_2 + 2 NH_3$

c) $C_3H_8 + 5 O_2 \longrightarrow 3 CO_2 + 4 H_2O$

35. a) $3 NO_2 + H_2O \longrightarrow 2 HNO_3 + NO$

b) $2 H_2S + SO_2 \longrightarrow 3 S + 2 H_2O$

c) $2 CH_3OH + 3 O_2 \longrightarrow 2 CO_2 + 4 H_2O$

37. a) $2 \text{ mol NH}_3 \times \dfrac{1 \text{ mol urea}}{2 \text{ mol NH}_3} = 1 \text{ mol urea}$

b) $0.45 \text{ mol NH}_3 \times \dfrac{1 \text{ mol urea}}{2 \text{ mol NH}_3} = 0.23 \text{ mol urea}$

c) $10 \text{ g NH}_3 \times \dfrac{1 \text{ mol NH}_3}{17.03 \text{ g}} \times \dfrac{1 \text{ mol urea}}{2 \text{ mol NH}_3} = 0.3 \text{ mol urea}$

d) $2.0 \text{ kg NH}_3 \times \dfrac{1000 \text{ g}}{1 \text{ kg}} \times \dfrac{1 \text{ mol NH}_3}{17.03 \text{ g}} \times \dfrac{1 \text{ mol urea}}{2 \text{ mol NH}_3} = 59 \text{ mol urea}$

39. $1.0 \times 10^3 \text{ kg SO}_3 \times \dfrac{1000 \text{ g}}{1 \text{ kg}} \times \dfrac{1 \text{ mol SO}_3}{80.06 \text{ g SO}_3} \times \dfrac{1 \text{ mol H}_2\text{SO}_4}{1 \text{ mol SO}_3} \times \dfrac{98.08 \text{ g H}_2\text{SO}_4}{1 \text{ mol H}_2\text{SO}_4} \times \dfrac{1 \text{ kg}}{1000 \text{ g}}$

$= 1.2 \times 10^3 \text{ kg H}_2\text{SO}_4$

SOLUTIONS TO FEATURE PROBLEMS:

45. a) CH_4

b) CO_2

c) $H_2N=NH_2$

d) H_2O_2

REVIEW TESTS FOR CHAPTER 4

Match the following statements or phrases to the end-of-chapter key terms.

1. Indicates the elements present in a compound and the relative number of atoms of each element.

2. A metal and a non-metal combine in a chemical reaction and form this type of bond.

3. These elements tend to lose their valence electrons to acquire full outer Bohr orbits.

4. Solution having dissolved ions.

5. Two non-metals combine in a chemical reaction to form this kind of bond.

6. Clusters of two or more atoms bonded together.

7. The sum of the atomic weights of all the atoms in a molecular formula.

8. Ions that contain more than one atom.

9. The substances on the lefts side of a chemical equation.

10. The substances on the right side of a chemical equation.

ANSWERS TO MATCHING

1. chemical formula
2. ionic
3. metals
4. electrolytes
5. covalent
6. molecules
7. molecular weight
8. polyatomic
9. reactants
10. products

SELF-TEST QUESTIONS

Completion: Write the word, phrase, or number in the blank that will complete the statement or answer the question.

1.   Calculate the molecular weight of CH_2Cl_2.

2.   Identify the product(s) in the following chemical reaction:

$$CaCO_3 \rightarrow CaO + CO_2$$

3.   Balance the following chemical reaction:

$$Al + H_2O \rightarrow Al(OH)_3 + H_2$$

4.   Name each of the following compounds:

    a. P_2O_5
    b. Na_3PO_4

5.   Give the chemical formulas for each of the following:

    a.   nitrogen trichloride
    b.   aluminum sulfate

6.   Calculate the number of methane molecules (CH_4) in 48.2 g of methane. $48.2g \times \dfrac{1 mole}{16 g}$

7.   How many moles of CO_2 are produced when 2.5 mol of O_2 react according to the following equation? Assume unlimited amounts of propane (C_3H_8),

$$C_3H_8 + 5\,O_2 \rightarrow 3\,CO_2 + 4\,H_2O$$

Multiple Choice: Select the correct answer from the choices listed.

8.   Which one of the following compounds is ionic?

    a) CO_2
    b) NH_3
    c)   NaF
    d) CCl_4

9.   Which one of the following compounds is covalent?

   a)   NaCl
   b) Na_2SO_4
   c) K_2O
   d) C_2H_6

10.   Which one of the following compounds contains a polyatomic ion?

   a) $NaNO_3$
   b)   KCl
   c)   MgO
   d) CO_2

11.   Which one of the following compounds is composed of discreet molecules?

   a)   KCl
   b) $MgBr_2$
   c) SO_2
   d) CaF_2

12.   Consider the following chemical formula. How many total oxygen atoms are present?
                     $Zn(NO_3)_2$

   a)   2
   b)   3
   c)   5
   d)   6

13.   Identify a typical property exhibited by ionic compounds.

   a)   form clusters of atoms called molecules
   b)   dissociate to form ions in solution
   c)   have directional bonds
   d)   are formed when a non-metal bonds with another non-metal

14.   What is the chemical formula for the compound formed between Ca and N?

   a) Ca_3N_2
   b) Ca_2N_3
   c) Ca_5N_2
   d)   CaN

15. What is the formula of the bicarbonate ion?

    a) OH^-
    b) CO_3^{2-}
    c) H_2CO_3
    d) HCO_3^-

16. What is the coefficient on H_2 when the following equation is correctly balanced?

    $$CO + H_2 \rightarrow H_2O + CH_4$$

    a) 1
    b) 2
    c) 3
    d) 4

17. How many molecules are in 1.0 kg of H_2O?

    a) 3.3×10^{25}
    b) 1.1×10^{25}
    c) 8.7×10^{23}
    d) 1.8×10^{27}

18. How many moles of carbon are in 2.5 moles of $C_6H_{12}O_6$?

    a) 0.42
    b) 15
    c) 1.5×10^{24}
    d) 9.0×10^{24}

ANSWERS TO SELF-TEST QUESTIONS

1. 84.93 g/mol
2. CaO, CO_2
3. $2\,Al + 6\,H_2O \rightarrow 2Al(OH)_3 + 3\,H_2$
4. a) diphosphorus pentoxide
   b) sodium phosphate
5. a) NCl_3
   b) $Al_2(SO_4)_3$
6. 1.81×10^{24} molecules
7. 1.5 mol CO_2
8. c
9. d
10. a
11. c

12. d
13. b
14. a
15. d
16. c
17. a
18. b

CHAPTER 5

THE BONDS THAT CHANGE

ANSWERS TO QUESTIONS:

1. As elements, both sodium and chlorine are very reactive due to their electron configurations. Their reactivity makes them harmful to biological systems. Sodium has one valence electron, while chlorine has seven valence electrons. Thus, both elements have incomplete octets. Sodium will lose one electron and chlorine will gain one electron to form a chemical bond. Both atoms are now stable due to octets in their outer shell. Therefore sodium and chloride when bonded together are relatively harmless.

3. Ionic bonding is represented by moving dots from the Lewis structure of the metal to the Lewis structure of the nonmetal to give both elements an octet. The dots surrounding the elements indicate the valence electrons. This corresponds to a stable configuration because the outermost occupied Bohr orbit contains an octet for each element. The metal and nonmetal each acquire a charge. Since opposite charges attract each other, there is an attractive force between the ions.

5. The Lewis theory is useful because it explains why elements combine in the observed ratios and allows one to predict the molecules that would form from certain elements. For example, fluorine, chlorine, bromine and iodine all exist as diatomic molecules in nature, just as predicted by the Lewis theory. Also, magnesium fluoride is composed of two fluoride ions and one magnesium ion, just as predicted by the Lewis theory.

7. **VSEPR** is an acronym for **valence shell electron pair repulsion**. This model or theory is useful in predicting the geometries of molecules formed from nonmetals. The main postulate of this model is that structure around a given atom is determined principally by minimizing electron-pair repulsions. The idea here is that bonding and nonbonding electron pairs around a given atom will be positioned as far apart as possible. This theory, in combination with the Lewis theory, can be used to predict the approximate molecular structure of a molecule.

9. A polar molecule has a slightly positive charge at one end and a slightly negative charge at the other end. The separation of charges is due to polar covalent bonds within the molecule. Polar bonds have two poles – a positive pole and a negative pole – due to a difference in electronegativities of the elements bonded. A molecule with polar bonds is polar unless the symmetry is such that the polar bonds cancel. The geometry as well as the polarity of its bonds determines whether a polyatomic molecule is polar. A nonpolar molecule arises from either nonpolar bonds or polar bonds that cancel each other out due to symmetry.

SOLUTIONS TO PROBLEMS:

11. a) $:\ddot{B}r\cdot$ b) $:\dot{\ddot{S}}\cdot$ c) $:\ddot{K}r:$ d) $He:$

   Kr and He are chemically stable

13. a) $[\,Li\,]^+ [:\ddot{F}:]^-$ b) $[\,Li\,]^+ [:\ddot{O}:]^{2-} [\,Li\,]^+$

   c) $[\,Sr\,]^{2+} [:\ddot{O}:]^{2-}$ d) $[:\ddot{I}:]^- [\,Sr\,]^{2+} [:\ddot{I}:]^-$

15. a) $[\,Na\,]^+ [:\ddot{O}:]^{2-} [\,Na\,]^+$ Na_2O

   b) $[:\ddot{S}:]^{2-} [\,Al\,]^{3+} [:\ddot{S}:]^{2-} [\,Al\,]^{3+} [:\ddot{S}:]^{2-}$ Al_2S_3

   c) $[:\ddot{C}l:]^- [\,Mg\,]^{2+} [:\ddot{C}l:]^-$ $MgCl_2$

   d) $[\,Be\,]^{2+} [:\ddot{O}:]^{2-}$ BeO

17. a) $:\ddot{F}—\ddot{O}—\ddot{F}:$

   b) $:\ddot{I}—N—\ddot{I}:$
      $|$
      $:\ddot{I}:$

   c) $:\ddot{S}=C=\ddot{S}:$

   d) $:\ddot{O}:$
        $\|$
      $:\ddot{C}l—C—\ddot{C}l:$

19. a) The Lewis structure how it is written is missing electrons which makes the end N and O and the O have incomplete octets. The correct Lewis structure is as follows:

   $:\ddot{N}=N=\ddot{O}:$

   b) O-S-O is incorrect because bonds are missing between the S and both oxygens, electrons are also missing which gives the atoms incomplete octets. The correct Lewis structure is as follows:

   $:\ddot{O}=\ddot{S}—\ddot{O}:$

38

c) Br-Br is incorrect because electrons are missing and both of the atoms have incomplete octets. The correct Lewis structure is as follows:

$$: \overset{..}{\underset{..}{Br}} - \overset{..}{\underset{..}{Br}} :$$

d) O=Si-O is incorrect because there is a bond missing between the Si and the last O, there is also electrons missing on the oxygens which give them incomplete octets. The correct Lewis structure is as follows:

$$: \overset{..}{\underset{..}{O}} = Si = \overset{..}{\underset{..}{O}} :$$

21. a) For OF_2, the total number of electron pairs is four, two bonding pairs and two nonbonding pairs. The electron geometry is tetrahedral, but since two of these electron pairs are lone pairs the resulting molecular geometry is bent.

b) For NI_3, the total number of electron pairs is four, three bonding pairs and one nonbonding pair. The electron geometry is tetrahedral, but since one of these electron pairs is a lone pair the resulting molecular geometry is pyramidal.

c) For CS_2, the total number of electron pairs is two (each double bond counts as one electron pair) and both pairs are bonding groups. The correct geometry is linear.

d) For Cl_2CO, the total number of electron pairs is three (carbon is the central atom), all pairs are bonding. The double bond between the carbon and oxygen counts as one electron pair. Therefore, the correct geometry is trigonal planar.

23. a) The total number of electron pairs around each individual carbon is two. Both pairs are bonding therefore both the electron and the molecular geometry is linear.

$$H - C \equiv C - H$$

b) The central atom is carbon in CCl_4; it has four electron pairs around it. All four electron pairs are bonding therefore both the electron and the molecular geometry is tetrahedral.

c) The central atom is P; it has four electron pairs around it. Three of the pairs are bonding and one is a lone pair. Therefore, the electron geometry is tetrahedral and the molecular geometry is pyramidal.

$$H-\overset{\cdot\cdot}{\underset{\underset{H}{|}}{P}}-H$$

d) The oxygens are the two central atoms, each with four electron pairs around it. Each oxygen has two bonding pairs and two lone pairs. Therefore, the electron geometry of each oxygen is tetrahedral and the molecular geometry is bent.

$$H-\overset{\cdot\cdot}{\underset{\cdot\cdot}{O}}\diagdown\underset{\cdot\cdot}{\overset{}{O}}\diagup H$$

25. a) NH_3 polar (pyramidal geometry)
   b) CCl_4 nonpolar (tetrahedral geometry – dipoles cancel each other)
   c) SO_2 polar (bent geometry)
   d) CH_4 nonpolar (tetrahedral geometry)
   e) CH_3OH polar (tetrahedral geometries at both the carbon and oxygen, making the molecule slightly bent at the oxygen)

SOLUTIONS TO FEATURE PROBLEMS:

33.

$$H-\overset{\overset{\displaystyle H}{|}}{\underset{\underset{\displaystyle :\overset{\cdot\cdot}{\underset{\cdot\cdot}{Cl}}:}{|}}{C}}\cdots\overset{\cdot\cdot}{\underset{\cdot\cdot}{Cl}}:$$

CH_2Cl_2 is a polar molecule with a slightly positive charge on the hydrogen side and a slightly negative charge on the chlorine side.

REVIEW TESTS FOR CHAPTER 5

Match the following statements or phrases to the end-of-chapter key terms.

1.   American chemist who developed a simple theory for chemical bonding.

2.   Electrons between two atoms.

3.   Electrons on a single atom.

4.   Theory which allows chemists to predict the shapes of molecules.

5.   Covalent bonds with uneven electron distributions.

6. The ability of an atom to attract electrons in a covalent bond.

7. Polar bonds can cancel in a molecule producing this type of molecule.

ANSWERS TO MATCHING

1. Lewis
2. bonding pair electrons
3. lone pair electrons
4. VSEPR theory
5. polar bonds
6. electronegativity
7. non-polar

SELF-TEST QUESTIONS

Completion: Answer each of the following questions.

1. Draw the electron dot structure for the element I.

2. Draw the Lewis structure for the following ionic compound: Na_2O

3. Draw the Lewis structure for the following covalent compound: NBr_3

4. Draw a Lewis structure and use the VSEPR model to determine the geometry for the following molecule: SF_2

5. Determine whether or not the following molecule is polar. CBr_4

6. Indicate what is wrong with the following Lewis structure, then correct the problem and write a correct Lewis structure.

$$\left[:\ddot{\underset{..}{Cl}}: \right]^{+} \quad \left[:\ddot{\underset{..}{O}}: \right]^{2-} \quad \left[:\ddot{\underset{..}{Cl}}: \right]^{+}$$

Multiple choice: Select the correct answer from the choices listed.

7. What is the molecular geometry predicted for CS_2?

   a) bent
   b) linear
   c) pyramidal
   d) trigonal planar

8.   What chemical formula does Lewis theory predict for the compound, potassium sulfide?

   a) KS_6
   b) K_6S
   c) KS_2
   d) K_2S

9.   Which one of the following molecules has polar bonds and yet is non-polar?

   a) F_2
   b) H_2S
   c)   LiCl
   d) SiF_4

10.   What is the total number of valence electrons contained in the following molecule?

$$S_2Cl_2$$

   a)   4
   b)   13
   c)   26
   d)   33

11.   Which one of the following elements will <u>not</u> contain an octet in its outermost Bohr orbit when it achieves a stable configuration?

   a)   F
   b)   Li
   c)   C
   d)   Mg

12.   Which one of the following compounds will have a lone pair of electrons on its central atom in its Lewis structure?

   a) BeH_2
   b) C_2H_6
   c) NCl_3
   d) CHF_3

13. Which electron dot structure is incorrect?

a) $\cdot \ddot{N} \vdots$ $^{3-}$

b) $\vdots \ddot{Br} \vdots$ $^{-}$

c) Ba^{2+}

d) H^{-}

14. What element on the periodic table has the greatest ability to attract an electron?

a) H (atomic #1)
b) He (atomic #2)
c) F (atomic #9)
d) Fr (atomic #87)

15. Select the correct charge distribution on the hydrogen iodide structure from the following list.

a) $\delta^+ \quad \delta^-$
   H – I

b) $\delta^- \quad \delta^-$
   H – I

c) $\delta^- \quad \delta^-$
   H – I

d) $\delta^+ \quad \delta^+$
   H – I

16. Which one of the following compounds contains one triple bond in its Lewis structure?

a) SO_3
b) HCN (carbon is central)
c) HOCl (oxygen is central)
d) I_2

17. A molecule with a tetrahedral electron geometry could have what possible molecular geometry(s)?

a) tetrahedral only
b) tetrahedral or pyramidal
c) tetrahedral, pyramidal, or bent

d) tetrahedral, trigonal planar, pyramidal
   or bent

18. A molecule with a trigonal planar electron geometry and a bent molecular geometry would contain how many bonding groups around the central atom?

   a) 1
   b) 2
   c) 3
   d) 4

ANSWERS TO SELF-TEST QUESTIONS

1.  $: \ddot{I} \cdot$

2. 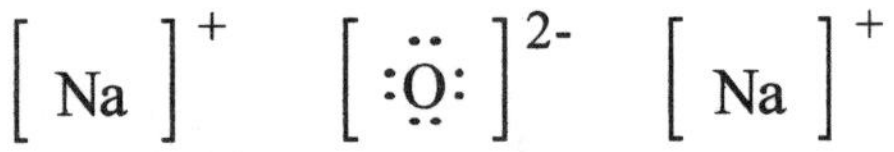 $\left[\text{Na} \right]^{+}$ $\left[: \ddot{O} : \right]^{2-}$ $\left[\text{Na} \right]^{+}$

3. 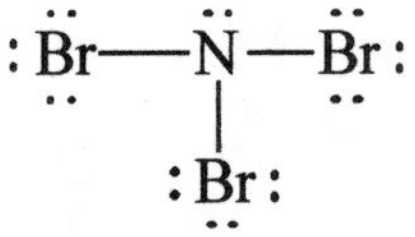 $: \ddot{Br} - \overset{\cdot\cdot}{N} - \ddot{Br} :$
        |
       $: \ddot{Br} :$

4. 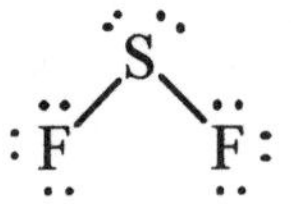 bent molecular geometry

5. Non-polar, since all the polar bonds cancel out. Lewis structure is shown below:

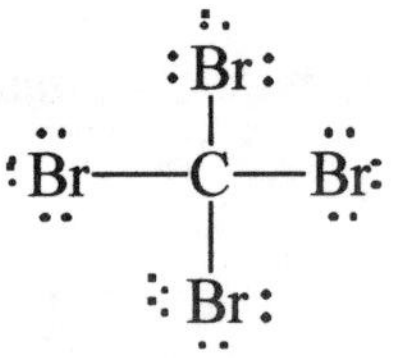

6. Structure is covalent, not ionic:

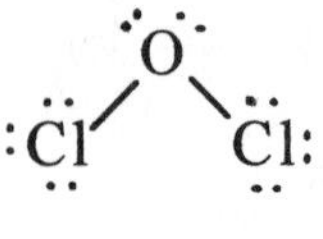

7. b
8. d

44

9.   d
10.   c
11.   b
12.   c
13.   a
14.   c
15.   a
16.   b
17.   c
18.   b

CHAPTER 6

ORGANIC CHEMISTRY

ANSWERS TO QUESTIONS:

1. Organic chemistry is the study of carbon-containing compounds.

3. a) table salt – inorganic
   b) sugar – organic
   c) copper - inorganic
   d) diamond – inorganic
   e) gold - inorganic
   f) vegetable oil – organic

5. Hydrocarbons are organic compounds that contain only carbon and hydrogen.

   alkane $C_n H_{2n+2}$
   alkene $C_n H_{2n}$
   alkyne $C_n H_{2n-2}$

7. Four common fuels are as follows:

   methane CH_4
   propane $CH_3CH_2CH_3$
   butane $CH_3CH_2CH_2CH_3$
   octane $CH_3(CH_2)_6CH_3$

9. Two important properties of alkanes are their flammability and nonpolar nature.

11. propene

   propyne

13. Isomers are molecules having the same molecular formula, but different structures.

15. a) aldehyde

$$R-\overset{\displaystyle O}{\overset{\|}{C}}H \qquad CH_3-\overset{\displaystyle O}{\overset{\|}{C}}H \qquad CH_3CH_2-\overset{\displaystyle O}{\overset{\|}{C}}H$$

b) ketone

$$R-\overset{\displaystyle O}{\overset{\|}{C}}-R \qquad CH_3-\overset{\displaystyle O}{\overset{\|}{C}}-CH_3 \qquad CH_3CH_2-\overset{\displaystyle O}{\overset{\|}{C}}-CH_3$$

c) carboxylic acid

$$R-\overset{\displaystyle O}{\overset{\|}{C}}-OH \qquad CH_3-\overset{\displaystyle O}{\overset{\|}{C}}-OH \qquad CH_3CH_2-\overset{\displaystyle O}{\overset{\|}{C}}-OH$$

d) ester

$$R-\overset{\displaystyle O}{\overset{\|}{C}}-OR \qquad CH_3-\overset{\displaystyle O}{\overset{\|}{C}}-OCH_3 \qquad CH_3CH_2-\overset{\displaystyle O}{\overset{\|}{C}}-OCH_3$$

e) ether

$$R-O-R \qquad CH_3-O-CH_3 \qquad CH_3CH_2-O-CH_3$$

f) amine

$$R-\overset{\displaystyle R}{\overset{|}{N}}-R \qquad CH_3-\overset{\displaystyle CH_3}{\overset{|}{N}}-H \qquad CH_3CH_2-\overset{\displaystyle CH_3}{\overset{|}{N}}-H$$

17. DDT, a chlorinated hydrocarbon, is an effective insecticide and is relatively nontoxic toward humans. However, many insects became resistant to DDT rendering it ineffective. DDT's excellent chemical stability became a liability. DDT became concentrated in the soil and eventually moved up the food chain killing fish and birds including the bald eagle.

19. Ethanol functions as a depressant on the central nervous system. Excessive alcohol consumption can lead to loss of coordination, unconsciousness and even death.

21. Due to its toxicity to bacteria, formaldehyde is used as a preservative of biological specimens.

23. Benzaldehyde is found in oil of almond, and cinnamaldehyde is found in cinnamon.

25. Formic acid is a component of the sting of biting ants, bees and wasps. Acetic acid is the active component in vinegar. Acetic acid gives vinegar and sourdough bread a bitter taste.

27. Ethyl butyrate is found in pineapples. Methyl butyrate is found in apples. Ethyl formate is found in artificial rum flavor. Benzyl acetate is a component of oil of jasmine.

SOLUTIONS TO PROBLEMS:

29.

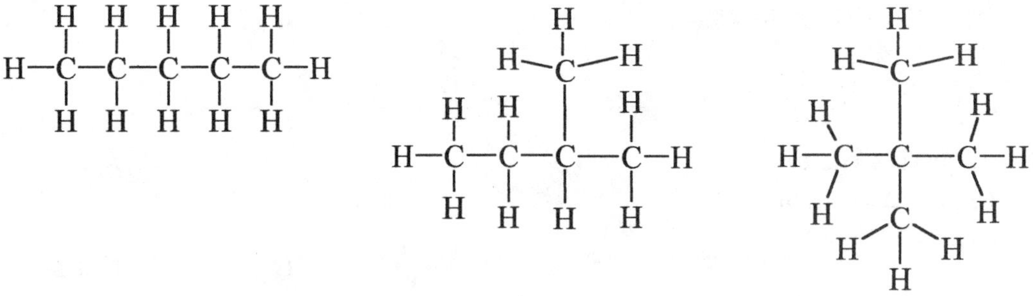

31.

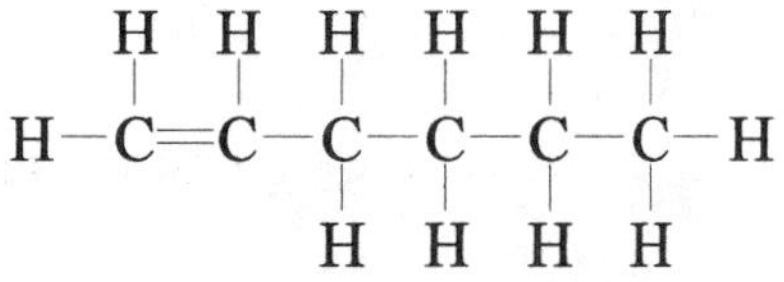

Two additional isomers are possible.

33. a) 2-methylbutane
    b) 4-ethyl-2-methylhexane
    c) 2,4-dimethylhexane
    d) 2,2-dimethylpentane

35. a) 2-butene
    b) 4-methyl-2-pentene
    c) 2-methyl-3-hexene

37. a) propyne
    b) 3-hexyne
    c) 4-ethyl-2-hexyne

39. a) $CH_3CHCH_2CH_2CH_3$ with CH_3 branch

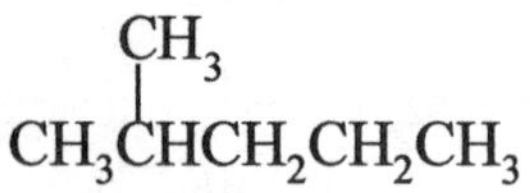

b) $CH_3CH_2CHCH_2CH_2CH_3$ with CH_3 branch

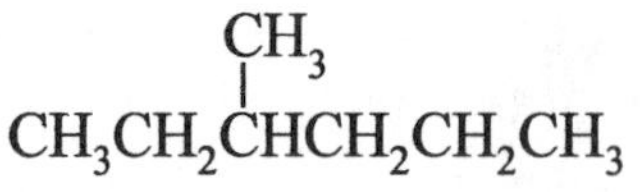

c) $CH_3CHCHCH_3$ with H_3C and CH_3 branches

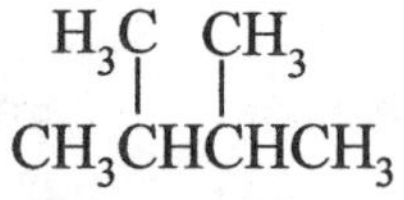

d) $CH_3CHCHCH_2CH_2CH_3$ with CH_3 branch and CH_2CH_3 branch

48

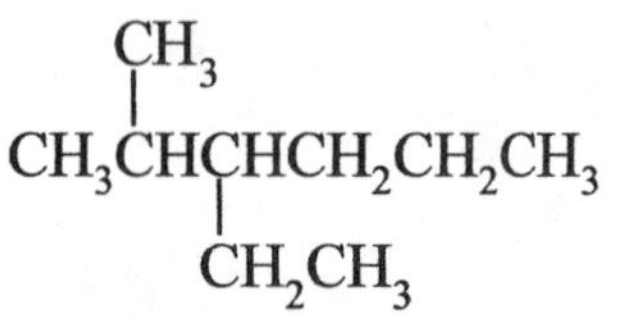

41. a) $CH_3CH{=}CHCH_3$ b) $CH_3CH_2C{\equiv}CCH_2CH_3$

c) 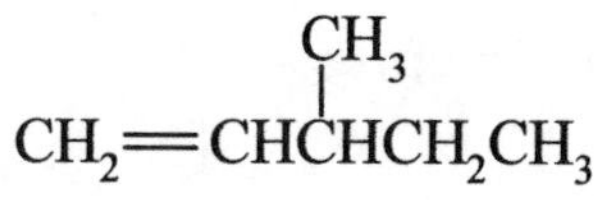
$$\underset{CH_2{=}CHCHCH_2CH_3}{\overset{\displaystyle CH_3}{\vert}}$$

d) 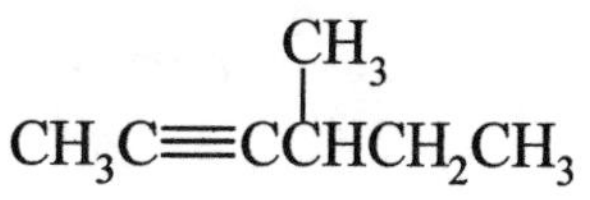
$$\underset{CH_3C{\equiv}CCHCH_2CH_3}{\overset{\displaystyle CH_3}{\vert}}$$

43. a) ether
    b) chlorinated hydrocarbon
    c) amine
    d) aldehyde

45. a) carboxylic acid
    b) aromatic hydrocarbon
    c) alcohol
    d) amine

47. Propanol is a polar molecule due to the presence of the OH functional group. Propane is a hydrocarbon, a nonpolar molecule. Polar molecules exhibit attractive intermolecular forces that tend to prevent the molecules from separating into the gas phase as easily as nonpolar molecules.

SOLUTIONS TO FEATURE PROBLEMS:

57. ethyl alcohol – CH_3CH_2OH
    water – H_2O
    diisopropylyamine –

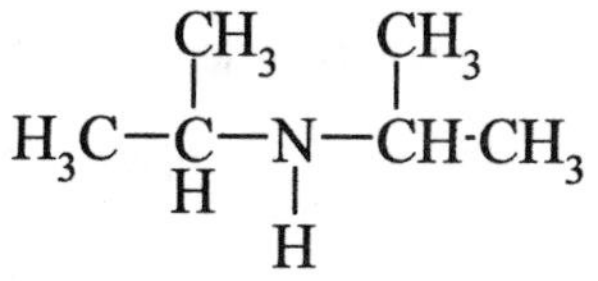

$$H_3C-\underset{H}{\overset{CH_3}{\underset{\vert}{\overset{\vert}{C}}}}-\underset{H}{\overset{\vert}{N}}-\overset{CH_3}{\overset{\vert}{CH}}\cdot CH_3$$

REVIEW TESTS FOR CHAPTER 6

Match the following statements or phrases to the end-of-chapter key terms.

1.   The study of carbon-containing compounds.

2.   The first organic compound synthesized from an inorganic compound in the laboratory.

3.   The family of hydrocarbons that contains only carbon and hydrogen.

4.   Name given to organic compounds that contain at least one double or triple bond.

5.   Molecules having the same molecular formula, but different structures.

6.   Benzene compounds contain these types of rings.

7.   A characteristic set of atoms that characterize a family of organic compounds.

8.   A molecule with many similar units bonded together in a long chain.

9.   Hydrocarbons can be systematically named according to guidelines established by this organization.

10.   The belief that living organisms could overcome ordinary physical laws and produce organic compounds.

ANSWERS TO MATCHING

1.   organic chemistry
2.   urea
3.   hydrocarbons
4.   unsaturated hydrocarbons
5.   isomers
6.   aromatic rings
7.   functional groups
8.   polymer
9.   I.U.P.A.C.
10.   vitalism

SELF-TEST QUESTIONS

Completion: Write the word, phrase, or number in the blank that will complete the statement or answer the question.

1. The general formula for alkynes is _____.

2. Write the condensed structural formula for butane.

3. Name the following organic compound:

$$CH_2=CHCH_3$$ ~~propylene~~

4. Write the condensed structural formula for the product formed in the following reaction:

$$CH_2=CH_2 + H_2 \rightarrow$$

5. The molecular formula for ethyne is _____.

6. Name each of the following organic compounds according to their IUPAC name:

a)
$$\begin{array}{c} CH_3 \\ | \\ CH_2 \\ | \\ CH_3CHCH_2CH_3 \end{array}$$

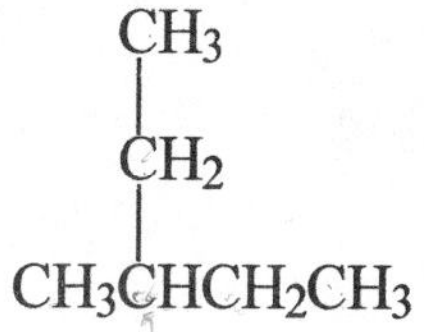

2 ethane butane (handwritten)

b)
$$\begin{array}{c} CH_3 \\ | \\ CH_3C=CHCH_2CH_3 \end{array}$$

2 methyl 2pentene (handwritten)

c)
$$\begin{array}{c} CH_3 \\ | \\ CH_3CH_2C{\equiv}CCHCH_2CH_3 \end{array}$$

3 ethyl 3 heptyne (handwritten)

51

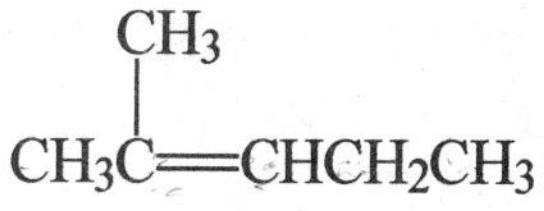

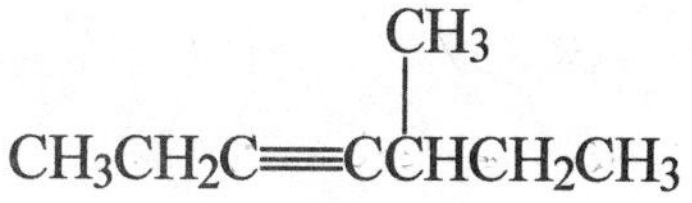

7.   Draw the condensed structural formula for each of the following organic compounds.

   a)   2,4-dimethyl-1-pentene

   b)   2-methyl propane

   c)   1-butyne

8.   An alcohol functional group is distinguished by having one or more _____ groups in the molecule.

9.   Draw the structural formulas for the two isomers of butane.

10. Amines are organic compounds that contain nitrogen and have the general formula _____.

Multiple choice: Select the correct answer from the choices listed.

11. What is the general formula for the aldehyde family?

   a)   RCOOH
   b)   RCHO
   c)   ROR
   d)   ROOR

12. What family of organic compounds typically exhibit fishy odors?

   a)   carboxylic acids
   b)   esters
   c)   alcohols
   d)   amines

13. What is the use of the chlorinated hydrocarbon, DDT?

   a)   as an insecticide
   b)   as a drug
   c)   as a refrigerant
   d)   as a preservative

14. What is the structure for benzene?

a)

b)

c)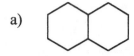

d)

15. How many hydrogen atoms are needed to complete the following structure?

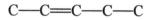   C—C=C—C—C

$h_2(s) = 10$

a)   8
b)   10
c)   12
d)   14

16. Which structure represents a ketone?

a)   CH₃CH₂——O——CH₂CH₃

b)

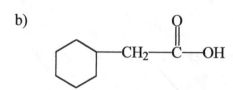

c)

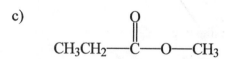

d)

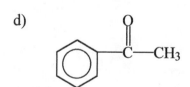

53

17. What is the structure for the methylene group?

   a) $-CH = CH -$

   b) $-CH_2 -$

   c) $CH_3 -$

   d) $CH_3CH_2 -$

18. What compound is responsible for the sweet aroma of pineapples?

   a) benzyl acetate
   b) acetic acid
   c) ethyl butyrate
   d) ethyl alcohol

19. Which one of the following carboxylic acids contains only one carbon atom?

   a) formic acid
   b) acetic acid
   c) citric acid
   d) lactic acid

20. Choose the substance that is classified as organic from the following list.

   a) graphite
   b) cobalt
   c) water
   d) cellulose

ANSWERS TO SELF-TEST QUESTIONS

1. C_NH_{2N-2}

2. $CH_3CH_2CH_2CH_3$

3. propene

4. CH_3CH_3

5. C_2H_2

6.   a)   3-methylpentane
       b)   2-methyl-2-pentene
       c)   5-methyl-3-heptyne

7.

                   CH$_3$ CH$_3$
                 | |
  a)   CH$_2$—CH—CH$_2$—CHCH$_3$

                 CH$_3$
                 |
  b)   CH$_3$CHCH$_3$

  c)   H—C≡C—CH$_2$CH$_3$

8.   – OH

9.

                                 CH$_3$
                                 |
  CH$_3$CH$_2$CH$_2$CH$_3$ and CH$_3$CHCH$_3$

10.   NR$_3$ (R = hydrogen or hydrocarbon group)

11.   b
12.   d
13.   a
14.   c
15.   b
16.   d
17.   b
18.   c
19.   a
20.   d

CHAPTER 7

LIGHT AND COLOR

ANSWERS TO QUESTIONS:

1. The colors present in white light are red, orange, yellow, green, blue, indigo and violet.

3. A magnetic field is the region around a magnet where forces are experienced. An electric field is the region around a charged particle where forces are experienced.

5. The wavelength of light determines its color. Wavelength also determines the amount of energy one of its photons carries. Wavelength and energy have an inverse relationship, indicating that as the wavelength increases, the energy decreases.

7. Sunscreens and UV-rated sunglasses contain compounds such as PABA that efficiently absorb ultraviolet light.

9. X-rays are a form of light that penetrates substances that normally block out light (opaque materials). X-rays are high-frequency, short-wavelength high-energy forms of electromagnetic radiation.

11. Night vision systems use infrared detectors to sense infrared light and to "see" in the dark. All warm objects, including human bodies, emit infrared light. Thus, if our eyes could respond to infrared light, people would glow like light bulbs. Microwave ovens work because microwaves are absorbed by water molecules causing them to heat up. Because most foods contain some water, microwaves are a quick and efficient way to heat food.

13. Spectroscopy is the interaction of light with matter. Spectroscopic methods of analysis are used by scientist to identify substances. This identification procedure is based on the wavelengths of light being absorbed or emitted by a particular substance.

15. In an MRI imaging apparatus, the sample to be studied is placed in an external magnetic field that is non-uniform in space. As a result, the nuclei of hydrogen atoms in the sample all experience a slightly different magnetic field. A sequence of radio frequencies is then applied to the sample, which causes the hydrogen nuclei to come into resonance (to flip). Each nucleus in the sample flips at a slightly different radio frequency, depending on the strength of the external magnetic field it is in. At every radio frequency that caused nuclei to flip, a line is plotted. The result is a spectrum showing an image of the object containing the sample (as shown in Figure 15). This

technique is incredibly useful for obtaining medical images since our bodies are abundant in hydrogen atoms.

17. A laser consists of three main components, a lasing medium sandwiched between two mirrors (one of which only partially reflects), and a laser cavity that encapsulates both the lasing medium and the mirrors. When a laser is turned on, molecules or atoms within the lasing medium are excited with light or electrical energy. The electrons in the molecules or atoms jump to higher energy states and when they relax back to their lower energy state they emit light (in the form of a photon). The photon released travels through the cavity, hits the mirror and bounces back through the lasing medium stimulating the emission of other photons, each with exactly the same wavelength and alignment as the original. The result is an amplification process that produces large numbers of photons circulating within the laser cavity. A small fraction of the photons leave the laser cavity through the partially reflecting mirror producing an intense, monochromatic laser beam.

19. The unique feature of the dye laser is its tunability. By choosing the proper dye and the correct laser cavity configuration, any wavelength of light in the visible region can be produced.

21. The use of a laser beam in medicine has several advantages over a scalpel. The advantages are listed below:

    a) Precise cuts through skin and tissue can be made with minimum damage to surrounding areas.
    b) The laser beam can be delivered through fiber-optic cable to difficult-to-reach places.
    c) The laser has a variable wavelength option that enables the surgeons to produce a variety of desirable effects.

SOLUTIONS TO PROBLEMS:

23. $1.5 \times 10^8 \text{ km} \times \dfrac{1000 \text{ m}}{1 \text{ km}} \times \dfrac{1 \text{ s}}{3.0 \times 10^8 \text{ m}} \times \dfrac{1 \text{ min}}{60 \text{ s}} = 8.3 \text{ min}$

25. The picture would have a rainbow of colors being absorbed by the red object and the color red being reflected by the object.

27. $\lambda = c/v$

$$\lambda = \frac{3.00 \times 10^8 \text{ m/s}}{93.6 \times 10^6 \text{ /s}} = 3.21 \text{ m}$$

29. $\lambda = c/\nu$

$$\lambda = \frac{3.00 \times 10^8 \text{ m/\cancel{s}}}{850 \times 10^6 \text{ /\cancel{s}}} = 0.353 \text{ m}$$

31. A lamp that emits light with a wavelength of 589 nm must contain sodium (Na).

33. Helium and hydrogen.

35. Blue light would be <u>reflected</u> and not absorbed by the tattoo. Any color other than blue
would be appropriate.

37. The sketch would look like Figure 6 on page 143 of the text. If that were a sketch of visible light, the UV light would have a shorter wavelength and infrared light would have a longer wavelength.

39. X-rays are the most energetic, while microwaves are the least energetic.

SOLUTIONS TO FEATURE PROBLEMS:

47. The problem with this picture is that the solution appears blue but the colors being reflected out the other side are red and orange. The color of the solution should correspond to the colors being reflected. Therefore, the solution should be red/orange.

REVIEW TESTS FOR CHAPTER 7

Match the following statements or phrases to the end-of-chapter terms.

1.   The green color of leaves is due to this class of molecules.

2.   Tiny packets of energy traveling at the speed of light.

3.   The area around a magnet where forces are experienced.

4.   The distance between wave crests in light's magnetic or electric fields.

5.   The number of cycles or crests that pass through a point in one second.

6. A general term for all forms of light.

7. Wilhelm Roentgen discovered this type of electromagnetic radiation.

8. The most energetic form of electromagnetic radiation.

9. Another name for the electron configuration of a molecule or atom with electrons in particular orbits.

10. The process in which an electron goes back to its original orbit, producing heat or light in the process.

11. The emission of light by the relaxation of excited electrons in which the object will glow in the dark.

12. The study of the interaction of light with matter.

13. The technology based on the spectroscopy of hydrogen atoms in a magnetic field.

14. The exact electromagnetic frequency that causes a transition.

15. A technology related to light that involves only one wavelength of light and whose crests are aligned or in phase.

ANSWERS TO MATCHING

1. chlorophylls
2. photons
3. magnetic field
4. wavelength
5. frequency
6. electromagnetic radiation
7. x-ray
8. gamma ray
9. energy state
10. relaxation
11. phosphorescence
12. spectroscopy
13. magnetic resonance imaging
14. resonant frequency
15. lasers

SELF-TEST QUESTIONS

Completion: Write the word, phrase, or number in the blank that will complete the statement or answer the question.

1. Carotenes, responsible for the color of carrots, absorb all colors except
    _____ and _____.

2. Light travels with a speed equal to _____ miles per second.

3. The unit of frequency is 1/second and is called the _____.

4. What color in the visible range would have an emission wavelength of 706 nm?

5. What is the wavelength in meters of radio waves produced by a radio station with a frequency of 1400 kHz?

6. Nuclear magnetic resonance involves the _____ of hydrogen atoms.

Multiple choice: Select the correct answer from the choices listed.

7. The scientist who used a glass prism to split light into its component colors would be

    a) Bloch
    b) Hertz
    c) Newton
    d) Roentgen

8. If the molecules in an object absorb all the light, there is no reflected light, and the object appears _____.

    a) red
    b) violet
    c) white
    d) black

9. As the wavelength of electromagnetic radiation decreases, the energy
    _____ and the frequency _____.

    a) increases, increases
    b) increases, decreases
    c) decreases, increases
    d) decreases, decreases

10. What is the wavelength range (in meters) for electromagnetic radiation?

    a) 4×10^{-7} to 7.8×10^{-7}
    b) 10^{23} to 10^4
    c) 10^{15} to 10^{-4}
    d) 10^{-15} to 10^4

11. All warm objects, including our own bodies, emit this type of electromagnetic
    radiation.

    a) gamma
    b) infrared
    c) visible
    d) microwave

12. What is the short-lived process in which electrons emit visible light in passing
    from a higher to a lower energy state?

    a) excitation
    b) phosphorescence
    c) fluorescence
    d) photodecomposition

13. MRI can measure the time required for nuclei to return to their original
    orientation after being pushed by electromagnetic radiation. What is this process
    called?

    a) relaxation time
    b) emission time
    c) resonant frequency
    d) absorption frequency

14. What type of laser is the most efficient and powerful and its beam can easily cut through steel?

   a)   diode
   b)   He – Ne
   c) CO_2
   d)   Nd: YAG

15. What characteristic of light determines both its color and energy?

   a)   speed
   b)   magnetic field strength
   c)   frequency
   d)   wavelength

16. In nuclear magnetic resonance technology, hydrogen nuclei act like tiny

   _____.

   a)   electrons
   b)   magnets
   c)   cells
   d)   photons

17. A ruby laser produces red light at 694 nm. How would this ruby laser be classified?

   a)   gas
   b)   dye
   c)   solid state
   d)   semi-conductor

18. Which one of the following forms at electromagnetic radiation has the shortest wavelength?

   a)   ultraviolet
   b)   x-ray
   c)   microwave
   d)   infrared

19. A red book seen in white light will _____ all the color components except the _____, which it _____.

    a) absorbs, red, reflects
    b) reflects, red, absorbs
    c) absorbs, white, absorbs
    d) reflects, blue, absorbs

ANSWERS TO SELF-TEST QUESTIONS

1. red, orange
2. 186,000
3. Hertz or Hz
4. red
5.

$$1400 \ \cancel{kHz} \times \frac{1000 \ Hz}{1 \ \cancel{kHz}} = 1.4 \times 10^6 \ Hz$$

$$\lambda = \frac{c}{\nu} = \frac{3.00 \times 10^8 \ \cancel{m/s}}{1.40 \times 10^6 \ \cancel{Hz}} = 2.1 \times 10^2 \ m$$

$$Hz = 1/s$$

6. nuclei
7. c
8. d
9. a
10. d
11. b
12. c
13. a
14. c
15. d
16. b
17. c
18. b
19. a

CHAPTER 8

NUCLEAR CHEMISTRY

ANSWERS TO QUESTIONS:

1. Radioactivity is the result of nuclear instability. An unstable nucleus will "decay" to achieve stability, releasing parts of its nucleus in the process. The parts released can collide with matter to produce large numbers of ions. Ions can damage biological molecules, which are contained in living organisms.

3. An alpha particle consists of two protons and two neutrons; it is a helium nucleus. A beta particle is an energetic, fast moving electron. A gamma ray is an energetic photon. Both alpha and beta particles are matter while gamma rays are electromagnetic radiation. Their ranking in terms of ionizing power is as follows:

$$\alpha \text{ particle} > \beta \text{ particle} > \Upsilon \text{ ray}$$

Their ranking in terms of penetrability is as follows:

$$\Upsilon \text{ ray} > \beta \text{ particle} > \alpha \text{ particle}$$

5. Radon gas is found in the soil near uranium deposits and also in the surrounding air. It represents a potential hazard because it is an unstable intermediate in the uranium radioactive decay series.

7. Albert Einstein wrote a letter to President Roosevelt describing the use of fission in the construction of very powerful bombs. American scientists were concerned that Nazi Germany would build a fission bomb first since that was where fission was discovered. President Roosevelt was convinced and decided that the United States must beat Germany to the bomb.

9. a) Los Alamos, New Mexico is the location where America's best scientists gathered to design the nuclear bomb.

   b) Hanford, Washington is the location where plutonium was manufactured.

   c) Oak Ridge, Tennessee is the place where uranium was processed.

   d) Alamogordo, New Mexico is the location where the first atomic bomb was tested.

11. During the fission reaction carried out by Fermi, a critical mass was achieved resulting in a self-sustaining reaction. Fission was occurring in a controlled manner. The fission reaction used to produce the atomic bomb is designed to escalate in an exponential manner to produce an explosion.

13. The scientist working on the Manhattan Project rationalized that the fission bomb would be better served for humanity in the United States versus Germany. The United States Government would scrutinize its control better than Nazi Germany.

15. Mass defect is the difference between the experimentally measured mass of a nucleus and the sum of the masses of the protons and neutrons in the nucleus. Nuclear binding energy is the energy related to the mass defect and represents the energy that holds a nucleus together.

17. "The China Syndrome" is a phrase used to describe the overheated core melting through the reactor floor and into the ground, (or all the way to China).

19. All nuclear waste is now being stored on the location where it was produced. The United States Government now has a program to begin building permanent disposal sites for the storage of nuclear waste. These underground storage facilities would keep the nuclear waste materials isolated from the public and the environment.

21. Fusion is a nuclear reaction in which nuclei of light atoms unite to form heavier nuclei. Since the heavier ones are more stable, energy is released in the process. The following nuclear equation describes a fusion reaction:

$$_1^2\text{H} + {}_1^3\text{H} \rightarrow {}_2^4\text{He} + {}_0^1\text{n}$$

23. External radioactivity can often be stopped by clothes on our bodies or by the skin. However, internal radioactivity is hazardous because the particles have direct access to vital organs.

25. Genetic defects have occurred in the offspring of laboratory animals upon exposure to high levels of radiation. However, the results of scientific studies of humans have not revealed an increase in genetic defects by radiation exposure. Few studies of humans exist because humans cannot purposely be exposed to radiation.

27. According to U-238 dating techniques, the earth is about 4.5 billion years old. To make the calculation, it is assumed no lead was originally present (rock was only uranium), that all of the $_{82}^{206}\text{Pb}$ formed over the years has remained in the rock, and the number of isotopes in the intermediate stages of decay between $_{92}^{238}\text{U}$ and $_{82}^{206}\text{Pb}$ is negligible. The last assumption is valid because once an $_{92}^{238}\text{U}$ isotope starts to decay

it reaches $^{206}_{82}$Pb relatively fast. The age according to U-238 dating is also considered valid since we can see stars today that are billions of light years away, because we can see these stars today, we know the earth must be billions of years old.

29. One way to obtain an image of a specific internal organ or tissue is to introduce a radioactive element into the patient. This radioactive element (as a part of a compound) will concentrate in the area of interest. Detectors outside the body can scan over a region and record the activity so that a complete activity image may be reconstructed.

SOLUTIONS TO PROBLEMS:

31.
a) $^{220}_{86}$Rn $\rightarrow$ $^{216}_{84}$Po $+$ $^{4}_{2}$He

b) $^{212}_{83}$Bi $\rightarrow$ $^{212}_{84}$Po $+$ $^{0}_{-1}$e

c) $^{224}_{88}$Ra $\rightarrow$ $^{220}_{86}$Rn $+$ $^{4}_{2}$He

d) $^{208}_{81}$Tl $\rightarrow$ $^{208}_{82}$Pb $+$ $^{0}_{-1}$e

33. $^{228}_{88}$Ra $\xrightarrow{\beta}$ $^{228}_{89}$Ac $\xrightarrow{\beta}$ $^{228}_{90}$Th $\xrightarrow{\alpha}$ $^{224}_{88}$Ra $\xrightarrow{\alpha}$ $^{220}_{86}$Rn

35. The number of 6 hour half-lives in 24 hours is 4, therefore the amount of Tc-99 left after 24 hours is:

20.0 μg / 2 = 10.0 μg after 1 half-life
10.0 μg / 2 = 5.00 μg after 2 half-lives
5.00 μg / 2 = 2.50 μg after 3 half-lives
2.50 μg / 2 = 1.25 μg after 4 half-lives

37. The total amount of Tc-99 that decayed in problem 35 is:

$$20 \text{ μg} - 1.25 \text{ μg} = 18.75 \text{ μg}$$

The total number of atoms decayed is:

$$18.75 \text{ μg} \times \frac{1 \text{ g}}{1 \times 10^6 \text{ μg}} \times \frac{1 \text{ mole}}{99 \text{ g}} \times \frac{6.022 \times 10^{23} \text{ atoms}}{1 \text{ mole}} = 1.141 \times 10^{17} \text{ atoms of Tc}$$

The total number of gamma rays emitted in 24 hours is:

$$1.141 \times 10^{17} \text{ atoms Tc} \times \frac{1 \text{ gamma ray}}{1 \text{ atom Tc - 99}} = 1.141 \times 10^{17} \text{ gamma rays emitted}$$

39. $^2_1H + ^2_1H \rightarrow ^3_2He + ^0_1n$

41. $\dfrac{78 \text{ mrem/yr}}{360 \text{ mrem/yr}} \times 100\% = 22\%$ of the exposure is due to medical X - rays

43. It is easiest to construct a table showing the amount of C-14 in the fossil as a function of time.

Number of Half-Lives	%C-14	Age of Fossil (yrs)
0	0.010	0
1	0.0050	5,730
2	0.0025	11,460
3	0.0012	17,190
4	0.0006	22,920

Because it takes 2 half-lives to reach a C-14 level of 0.0025, the fossil must be 11,460 years old.

45. Since a living organism has approximately .010% C-14 we must take 6.25 % of that to find out the percentage of C-14 in the skeleton.

$.0625 \times .010\% \text{ C - 14} = .000625\% \text{ C - 14}$ in the skeleton

Using the table from the answer of number 43, we see that the fossil must be just slightly older than 22,920 years old.

47. Using the table from problem 46 it is easy to see that the rock containing 50% uranium and 50% lead must be 4.5×10^9 years old.

SOLUTIONS TO FEATURE PROBLEMS:

59. The missing particles would consist of 5 reds and 7 grays.

$^{16}_7N \rightarrow ^{12}_5B + ^4_2He$

REVIEW TESTS FOR CHAPTER 8

Match the following statements or phrases to the end-of-chapter terms.

1.   Scientist who discovered the elements radium and polonium.

2.   An energetic electron emitted by an atomic nucleus.

3.   An energetic photon emitted by an atomic nucleus.

4.   The time required for half the nuclei in a radioactive sample to decay.

5.   The process of combining lighter nuclei to form heavier ones.

6.   The difference between the experimentally measured mass of a nucleus and the sum of its masses of the protons and neutrons.

7.   The energy that holds a nucleus together.

8.   The minimum mass of radioactive material that will sustain a nuclear chain reaction.

9.   The scientist who was the director of the Manhattan Project.

10.   The emission of energetic particles by an unstable nucleus.

ANSWERS TO MATCHING

1.   Marie Curie
2.   beta radiation
3.   gamma radiation
4.   half-life
5.   nuclear fusion
6.   mass defect
7.   nuclear binding energy
8.   critical mass
9.   J. R. Oppenheimer
10.   radioactivity

SELF-TEST QUESTIONS

Completion: Write the word, phrase, or number in the blank which will complete the statement or answer the question.

1.   Write the nuclear equation for the beta decay of P-32.

$$^{32}_{15}P \rightarrow ^{32}_{16}S + ^{0}_{-1}e$$

2.   Write a nuclear equation to describe a fusion reaction.

3.   The most common unit for measuring human exposure to radiation is called the
      _____.

4.   The half-life of Pr-145 is 6 hours. If you had 40 grams of Pr-145 at 1:00 P.M., how much would remain the next day at 1:00 P.M.?

5.   The C-14 in an ancient bow is found to be 25% of that found in living organisms. What is the age of the bow?

6.   What percent of the total human radiation exposure is due to natural sources?

Multiple choice: Select the correct answer from the choices listed.

7.   If a person was exposed to a radiation dose of 50 rem, what would be a probable outcome?

    a)   radiation sickness
    b)   lesions on skin
    c)   decrease in white blood cell count
    d)   death

8.   What is the greatest single source of radiation exposure for humans?

    a)   medical x-ray
    b)   radon
    c)   cosmic rays
    d)   nuclear medicine

9. What is the basis for the C-14 dating method?

    a) C-14 is very unstable and is readily lost from the atmosphere.
    b) All living organisms contain a residual amount of C-14.
    c) Living organisms will not absorb C-14 but will absorb C-12.
    d) When an organism dies, it continues to take in C-14.

10. What does the curie measure?

    a) total energy absorbed by an object exposed to a radioactive source.
    b) lethal threshold for radiation exposure.
    c) number of alpha particles emitted by one gram of a radioactive substance.
    d) number of disintegrations per second of a radioactive substance.

11. An alpha particle is represented by what symbol?

    a) $_{-1}^{0}e$

    b) $_{1}^{0}e$

    c) $_{2}^{4}He^{2+}$

    d) $_{1}^{2}H$

12. What is a major problem in exploiting nuclear fusion as an energy source?

    a) the high radioactivity of the reactants.
    b) the high temperatures required for the reaction.
    c) the lack of the appropriate reactants.
    d) the reaction is difficult to control.

13. Complete the following conversion factor which is used to measure air-borne radon.

$$1 \text{ pCi} = ? \text{ Ci}$$

    a) 10^{-12}
    b) 10^{12}
    c) 10^{-9}
    d) 3.7×10^{10}

14. The percentages of uranium and lead in a meteorite were 70% and 30%, respectively. How old is the meteorite?

a) exactly 4.5×10^9 yrs.
b) exactly 9.0×10^9 yrs.
c) older than 9.0×10^9 yrs.
d) younger than 4.5×10^9 yrs.

15. What type of radiation has the highest penetrating power?

a) α
b) β
c) γ

16. This reaction is an example of _____ decay.

$$^{204}_{82}Pb \longrightarrow {}^{204}_{82}Pb + \underline{\hspace{2cm}}$$

a) alpha
b) beta
c) gamma
d) neutron

17. What newly discovered man-made element was used during the Manhattan Project to produce additional uranium-235?

a) polonium – 210
b) plutonium – 239
c) thorium – 234
d) plutonium – 244

18. A nucleon refers to what subatomic particle(s)?

a) proton only
b) neutron only
c) both proton and neutron
d) proton and neutron

19. Elements with an atomic number near _____ have the most
_____ nuclei.

    a)   26, stable
    b)   26, unstable
    c)   83, stable
    d)   83, unstable

20. What is the role of the cadmium rods in a nuclear reactor?

    a)   cool the reactor core
    b)   produce more neutrons
    c)   provide fuel to the core
    d)   control the rate of reaction in the core.

ANSWERS TO SELF-TEST QUESTIONS

1. $^{32}_{15}P \longrightarrow \, ^{32}_{16}S \, + \, ^{0}_{-1}e$

2. $^{2}_{1}H \, + \, ^{3}_{1}H \longrightarrow \, ^{4}_{2}He \, + \, ^{1}_{0}N$

3.   rem
4.   2.5 g
5. 9.0×10^{9} yrs.
6.   82%
7.   c
8.   b
9.   b
10.   d
11.   c
12.   b
13.   a
14.   d
15.   c
16.   c
17.   b
18.   d
19.   a
20.   d

CHAPTER 9

ENERGY FOR TODAY

ANSWERS TO QUESTIONS:

1. The hotter an object is, the faster its molecules move. The cooler an object is, the slower its molecules move.

3. Heat is the random motion of molecules or atoms. Work is the use of energy to move atoms and molecules in a non-random or orderly fashion.

5. The U.S. consumes approximately 94 quads of energy per year.

7. The second law of thermodynamics states that for any spontaneous process, the universe, the system and the surroundings must become disorderly. Another word for disorder is entropy. Thus another way to state the second law is that for any spontaneous process entropy must increase. Because of the second law, a chemical reaction that gives off energy must release some as heat in order to increase the entropy of the universe. As a result, all the energy given off by a chemical reaction is not completely utilized.

9. a) Gasoline energy to forward motion of an automobile – 20% efficient.
    b) Food energy to physical work – 45% efficient.
    c) Fossil fuel energy to electricity – 25-40% efficient.
    d) Natural gas energy to heart – 70-90% efficient.

11. Power is not energy, but rather determines how long it will take to use a given amount of energy.

13. Heat capacity of a substance is the amount of heat energy required to change the temperature of a given amount of the substance by 1°C. A substance with a large heat capacity absorbs a lot of heat without a large increase in temperature. A low heat capacity substance cannot absorb a lot of heat without a large increase in temperature.

15.

Fuel	Years Until Depletion
Petroleum	60
Natural Gas	120
Coal	1500
Nuclear	100
Hydroelectric	indefinite
Other	

17. Fossil fuel burning power plants use heat released in combustion reactions to boil water, generating steam that turns the turbine of an electric generator. The generator is then used to create electricity that is transmitted to buildings via power lines. A typical fossil fuel plant produces 1 gigawatt (1 giga = 10^9) of power in the form of electricity, which can light 1 million homes.

19. Nitrogen dioxide (NO_2) – eye and lung irritant

Ozone (O_3) and PAN ($CH_3CO_2NO_2$) – lung irritant, difficulty in breathing, eye irritant,
damages rubber products, and damages crops.

Carbon monoxide (CO) – toxic because it diminishes the blood's capacity to carry oxygen.

21. Acid rain is caused by the emission of sulfur dioxide (SO_2) that results from fossil fuel combustion especially coal-burning power plants. The emission of nitrogen monoxide (NO) and nitrogen dioxide (NO_2) also contribute to acid rain.

23. The earth's atmosphere is transparent to visible light from the sun. This visible light strikes the earth, and part of it is changed to infrared radiation. This infrared radiation from the earth's surface is strongly absorbed by CO_2 and H_2O molecules in the atmosphere. In effect, the atmosphere traps some of the energy, acting like the glass in a greenhouse and keeping the earth warmer than it would otherwise be.

25. The burning of fossil fuels by human beings has increased the amount of carbon dioxide in the earth's atmosphere. This leads to the warming of the earth hence the term global warming. Based on scientific data, the global mean temperature has increased over the last century. Thus, global warming has been occurring.

27. Carbon dioxide levels have increased about 20% over the last century. The temperature has increased approximately 1°F corresponding to this increase in CO_2 concentration.

SOLUTIONS TO PROBLEMS:

29. a) $1456 \text{ cal} \times \dfrac{1 \text{ Cal}}{1000 \text{ cal}} = 1.456 \text{ Cal}$

b) $450 \text{ cal} \times \dfrac{4.18 \text{ J}}{1 \text{ cal}} = 1.88 \times 10^3 \text{ J}$

74

c) $20 \, kWh \times \dfrac{3.6 \times 10^6 \, J}{1 \, kWh} \times \dfrac{1 \, cal}{4.18 \, J} = 1.7 \times 10^7 \, cal$

d) $84 \, quads \times \dfrac{1.06 \times 10^{18} \, J}{1 \, quad} = 8.9 \times 10^{19} \, J$

31. a) $5/9 \, (212 \, °F - 32) = 100°C$
    b) First convert K to °C

$$77 \, K \quad = °C + 273$$
$$\underline{-273} \qquad \underline{-273}$$
$$-196 \quad = °C$$

then convert °C to °F
$°F = 9/5 \, (-196) + 32$
$°F = -321$

c) $K = 25 + 273$
    $K = 298$

d) First convert °F to °C
$°C = 5/9 \, (100 - 32) = 37.8 \, °C$

then convert °C to K
$K = 37.8 + 273$
$K = 311$

33. $°C = 5/9 \, (-80 - 32) = -62 \, °C$
    $K = -62 \, °C + 273$
    $K = 211 \, K$

35. $200 \, kcal \times \dfrac{1000 \, cal}{1 \, kcal} \times \dfrac{4.18 \, J}{1 \, cal} \times \dfrac{1 \, kWh}{3.6 \times 10^6 \, J} = .23 \, kWh$

$.23 \, kWh \times \dfrac{1 \, h}{0.10 \, kWh} \times \dfrac{60 \, minutes}{1 \, h} = 138 \, min$

Use conversion factors in Table 3 and Table 4.

37. a) $100 \text{ W} \times \dfrac{1 \text{ kW}}{1000 \text{ W}} = 0.1 \text{ kW}$

$\dfrac{5 \text{ hours}}{1 \text{ day}} \times \dfrac{30 \text{ days}}{1 \text{ month}} = 150$ hours per month

$0.1 \text{ kW} \times 150 \text{ hours} = 15 \text{ kWh}$

$15 \text{ kWh} \times \dfrac{\$0.10}{\text{kWh}} = \$1.50$

b) $600 \text{ W} \times \dfrac{1 \text{ kW}}{1000 \text{ W}} = 0.6 \text{ kW}$

$\dfrac{24 \text{ hours}}{1 \text{ day}} \times \dfrac{30 \text{ days}}{1 \text{ month}} = 720$ hours per month

$0.6 \text{ kW} \times 720 \text{ hours} = 432 \text{ kWh}$

$432 \text{ kWh} \times \dfrac{\$0.10}{\text{kWh}} = \$43.20$

c) $12{,}000 \text{ W} \times \dfrac{1 \text{ kW}}{1000 \text{ W}} = 12 \text{ kW}$

$\dfrac{1 \text{ hour}}{1 \text{ day}} \times \dfrac{30 \text{ days}}{1 \text{ month}} = 30$ hours per month

$12 \text{ kW} \times 30 \text{ hours} = 360 \text{ kWh}$

$360 \text{ kWh} \times \dfrac{\$0.10}{\text{kWh}} = \$36.00$

d) $1{,}000 \text{ W} \times \dfrac{1 \text{ kW}}{1000 \text{ W}} = 1 \text{ kW}$

$\dfrac{10 \text{ minutes}}{60 \text{ minutes}} \times \dfrac{1 \text{ hour}}{1 \text{ day}} \times \dfrac{30 \text{ days}}{1 \text{ month}} = 5$ hours per month

$1 \text{ kW} \times 5 \text{ hours} = 5 \text{ kWh}$

$$5 \text{ kWh} \times \frac{\$0.10}{\text{kWh}} = \$.50$$

39. a) Pinewood:

$$50 \text{kg} \times \frac{1000 \text{ g}}{1 \text{kg}} \times \frac{-5.1 \text{kcal}}{1 \text{g}} = -2.6 \times 10^5 \text{ kcal}$$

(negative sign indicates energy is emitted)

b) Coal:

$$2000 \text{ kg} \times \frac{1000 \text{ g}}{1 \text{kg}} \times \frac{-6.8 \text{kcal}}{1 \text{g}} = -1.4 \times 10^7 \text{ kcal}$$

c) Gasoline (isooctane):

$$60 \text{ L} \times \frac{1000 \text{ mL}}{1 \text{L}} \times \frac{0.7028 \text{ g}}{1 \text{mL}} \times \frac{-8.7 \text{kcal}}{1 \text{g}} = -3.7 \times 10^5 \text{ kcal}$$

41. a) 200 kcal x 0.3 = 60 kcal
    b) Refer to figure 2.
       2200 kcal x 0.45 = 990 kcal
    c) $\dfrac{1000 \text{ kcal}}{0.20} = 5000 \text{ kcal}$

    d) $\dfrac{1000 \text{ kcal}}{0.34} = (2941 \text{ kcal}) = 3000 \text{ kcal}$

43. a) $\dfrac{1000 \text{ kcal}}{0.80} = 1250 \text{ kcal}$

    $$-1250 \text{ kcal} \times \frac{1 \text{g}}{-11.8 \text{kcal}} = 1.1 \times 10^2 \text{ g natural gas}$$

    b) $\dfrac{1250 \text{ kcal}}{0.30} = 4170 \text{ kcal}$

    $$-4170 \text{ kcal} \times \frac{1 \text{g}}{-6.8 \text{kcal}} = 6.1 \times 10^2 \text{ g coal}$$

45. This is a stoichiometry problem

$$CH_4 + 2O_2 \xrightarrow{\Delta} 2H_2O + CO_2$$

a) From the balanced combustion equation above, 1 mol of methane (natural gas) produces 1 mol of carbon dioxide gas (CO_2).

Use answer in 43a. to determine the number of grams of CO_2 produced.

$$106\,g\,CH_4 \times \frac{1\,mol\,CH_4}{16.01\,g\,CH_4} \times \frac{1\,mol\,CO_2}{1\,mol\,CH_4} \times \frac{44.01\,g\,CO_2}{1\,mol\,CO_2} = 2.9 \times 10^2\,g\,CO_2$$

b) Use 4170 kcal of coal from 43b.

$$4170\,kcal \times \frac{1\,gram\,CO_2}{1.25\,kcal} = 3.3 \times 10^3\,g\,CO_2$$

47. a) $2C_8H_{18} + 25O_2 \rightarrow 16CO_2 + 18H_2O$

b) $150\,gal \times \dfrac{3.78\,L}{1\,gal} \times \dfrac{1000\,mL}{1\,L} \times \dfrac{.79\,g}{1\,mL} \times \dfrac{1\,mol}{114\,g} = 393\,moles\,C_8H_{18}$

c) $393\,moles\,C_8H_{18} \times \dfrac{16\,moles\,CO_2}{2\,moles\,C_8H_{18}} \times \dfrac{44\,g\,CO_2}{1\,mol} \times \dfrac{1\,kg}{1000g} = 138\,kg$

SOLUTIONS TO FEATURE PROBLEMS:

55. a)
   1998   90 Quadrillion
   1959   -37 Quadrillion
             53 Quadrillion increase

$$yearly\ avg.\ increase = \frac{53\,Quad}{39\,yrs} = 1.4\,Quad/yr$$

b) 70 quad – 37 quad = 33 quad
   33 quad / 39 quad = 0.85 quad/ year

c) 70 quad – 90 quad = -20 quad

d) 2020 −1998 = 22 years

22 x 1.4 + 90 = 120.8 quads consumed
22 x 0.85 + 70 = 88.7 quads produced

88.7quad − 120.8 quad = -32.1 quad

REVIEW TESTS FOR CHAPTER 9

Match the following statements or phrases to the end-of-chapter key terms.

1.   Random molecular or atomic motion.

2.   The study of energy and its transformation from one form to another.

3.   Energy can neither be created nor destroyed, only transferred between the system and its surroundings.

4.   The measure of disorder in the universe.

5.   A force acting through a distance.

6.   The rate of energy output or input.

7.   The temperature scale in which water boils at 100°.

8.   The term used for chemical reactions that give off energy to the surroundings.

9.   The fossil fuel composed of a mixture of methane and ethane.

10.   The term given to a reaction which is the reverse of photosynthesis.

11.   Ozone and PAN are the main components of this air pollutant.

12.   Substances that promote chemical reactions without being consumed.

13.   A specific term given to gases that allow visible light into the atmosphere, but prevent heat in the form of infrared light from escaping.

14.   The capacity to do work.

15. The temperature scale that does not contain negative values.

16. The amount of heat absorbed or emitted by a chemical reaction.

ANSWERS TO MATCHING

1. heat
2. thermodynamics
3. first law of thermodynamics
4. entropy
5. work
6. power
7. Celsius
8. endothermic
9. natural gas
10. combustion
11. photochemical smog
12. catalysts
13. greenhouse gases
14. energy
15. Kelvin
16. enthalpy of reaction

SELF-TEST QUESTIONS

Completion: Write the word, phrase, or number in the blank which will complete the statement or answer the question.

1. The hotter the object is, the _____ it molecular motion.

2. If we examine a chemical reaction occurring in a test tube, the system is _____.

3. Power × Time = _____.

4. The higher the energy output per gram of CO_2, the better the fuel with respect to global warming because _____ CO_2 will be emitted in producing a given amount of energy.

5. Write a balanced chemical equation for the combustion of pentane (C_5H_{12}).

6.	Given H_{rxn} = -6.8 kcal/g for coal, calculate the amount of energy in kcal produced by the combustion of 5.2×10^3 kg of coal.

Multiple choice: Select the correct answer from the choices listed.

7.	Which one of the following involved a decrease in entropy?

    a)	dissolving sugar in tea
    b)	putting together a crossword puzzle
    c)	making a chocolate milkshake
    d)	blowing up a balloon

8.	Of the following units, which one is <u>not</u> a unit of energy?

    a)	Joule
    b)	Calorie
    c)	Watt
    d)	Quad

9.	What is the temperature at which all molecular motion stops?

    a)	100 K
    b)	273 K
    c)	-273° F
    d)	-273° C

10.	What is the chemical formula for the brown gas that gives smog its characteristic color?

    a)	NO_2
    b)	NO
    c)	CO_2
    d)	CO

11.	What type of device violates the first law of thermodynamics?

    a)	battery
    b)	windmill
    c)	x-ray machine
    d)	perpetual motion machine

12. Which one of the following processes would have a negative enthalpy of reaction?

   a) the burning of gasoline
   b) a chemical cold pack
   c) the melting of ice
   d) the boiling of water

13. What factor would not contribute to increased global warming?

   a) fluctuations in the Sun's radiation
   b) decrease in the amount of CO_2 in the atmosphere
   c) changing ocean currents
   d) deforestation

14. What energy-containing molecule used by plants and animals is produced during photosynthesis?

   a) water
   b) methane
   c) glucose
   d) carbon dioxide

15. Water has one of the highest heat capacities known. What does this mean with respect to the relationship of heat to changes in temperature?

   a) Water can absorb a lot of heat with a large temperature change.
   b) Water can absorb a lot of heat without a large increase in temperature.
   c) Water can absorb a lot of heat with no temperature change.
   d) Water can absorb very little heat with a large increase in temperature.

16. The major source of CO pollution is

   a) industry
   b) domestic heating
   c) transportation
   d) combustion for electrical generation

17. When SO_2 and NO_2 combine with water they form acids that produce which ions.

   a) H^+
   b) OH^-
   c) O^{2-}
   d) H^-

18. Which one of the following is <u>not</u> a fossil fuel?

   a) coal
   b) ozone
   c) natural gas
   d) petroleum

19. The pollutant primarily responsible for acid rain is

   a) NO_x
   b) CO_2
   c) SO_2
   d) O_3

20. According to the second law of thermodynamics, what is the relationship of the spontaneity of chemical reactions and the order of the universe?

   a) Spontaneous chemical reactions must increase the disorder of the universe.
   b) Spontaneous chemical reactions must decrease the disorder of the universe.
   c) Nonspontaneous chemical reactions must increase the disorder of the universe.
   d) No relationship exists since the temperature at which a chemical reaction occurs can vary.

ANSWERS TO SELF-TEST QUESTIONS

1. faster
2. chemicals
3. energy
4. less
5. $C_5H_{12} + 8 O_2 \rightarrow 5 CO_2 + 6 H_2O$

6.

$$5200 \, \cancel{kg} \times \frac{1000 \, \cancel{g}}{1 \, \cancel{kg}} \times \frac{-6.8 \, kcal}{\cancel{g}} = -3.5 \times 10^7 \, kcal$$

7.   b
8.   c
9.   d
10. a
11. d
12. a
13. b
14. c
15. b
16. c
17. a
18. b
19. c
20. a

CHAPTER 10

ENERGY FOR TOMORROW: SOLAR AND OTHER RENEWABLE ENERGY SOURCES

ANSWERS TO QUESTIONS:

1. The main obstacle to using solar energy is its low concentration

3. As water flows through a dam, it turns a turbine on a generator that creates electricity.

5. The winds turns the blades on the turbine which they generates electricity.

7. A solar power tower allows for the generation of electricity for several hours in the dark. The sun's rays are focused onto a central receiver located on top of a tower. The high temperatures produced are used to heat a molten-salt liquid that circulates into a storage tank. The molten salt is pumped out of the storage tank, on demand, to generate steam, which then turns a turbine on an electrical generator.

9. The disadvantages of solar thermal technology are the high maintenance costs, the low concentration, and uncontrollable weather conditions. The advantages of solar thermal technology are its efficiency, renewability and lack of any air pollution.

11. A semiconductor is a material that conducts electricity under some conditions but not others. The n-type silicon semi-conductor stands for negative. The p-type silicon semi-conductor stands for positive.

13. Photovoltaic cells are quite expensive. Semi-conductors are expensive to manufacture and photovoltaic cells are costly to produce. Also, photovoltaic cells are inefficient with respect to converting incident energy into electricity. On the advantageous side, photovoltaic cells contain no moving parts, produce no noise and are environmentally safe.

15. One source of biomass energy is in the molecules of plants. Biomass energy can be transported easily, as in liquid ethanol and burns cleanly. Also, there is little effect on global warming since the balance is equal between the amounts of CO_2 released when the plant is burned for energy and the CO_2 absorbed to make the biomass. The major problem with biomass energy is the amount of cropland required for it to become a significant energy source.

17. A nuclear power plant is essentially a steam engine using uranium as a fuel. It suffers from low efficiency characteristic of all heat engines and the accompanying thermal

pollution. Nuclear power plants produce practically no air pollution. However, accidents can release high levels of damaging radioactivity. Nuclear power plants present other problems as well: radioactive substances produced are difficult to dispose of; and the nuclear fuel supply is limited. But the energy produced per kilogram of fuel is very large and extraction is less damaging to the land than for fossil fuels. The public's fear of accidents and concerns over waste disposal has suppressed the use of nuclear power in the U.S.

19. Instead of using an automobile for transportation, riding a bike, ride sharing, walking or using public transportation could be substituted. Thus, energy could be conserved. Conserving energy in households involves improved methods of insulation, the lowering of thermostats and using extra sweaters or blankets.

21. It is clear that conserving our limited fuel supplies, by avoiding wasteful use of energy, should be of great concern to our society. The work involved with unleashing greater amounts of energy and with cleaning up environmental pollution would increase energy costs.

SOLUTIONS TO PROBLEMS:

23.
$$\$195.00 \times \frac{1 \text{kWh}}{\$0.12} = 1625 \text{ kWh}$$

$$1625 \text{ kWh} \times \frac{\$0.16}{\text{kWh}} = \$260$$

25.
$$\$245 \times \frac{1 \text{ kWh}}{\$0.11} = 2{,}227 \text{ kWh}$$

$$2{,}227 \text{ kWh} \times \frac{3.6 \times 10^6 \text{ J}}{1 \text{ kWh}} = 8.0 \times 10^9 \text{ J}$$

27. % efficiency $= \dfrac{71 \text{ W}}{568 \text{ W}} \times 100\% = 13\%$

29. $1487 \text{ W} \times 0.16 = 238 \text{ W}$ of power produced by the PV cell

31.
$$\text{power in} = \frac{5.0 \text{ kW}}{0.18} = 27.8 \text{ kW}$$

$$27.8 \text{ kW} \times \frac{1000 \text{ W}}{1 \text{ kW}} \times \frac{\text{m}^2}{1000 \text{ W}} = (27.8 \text{ m}^2) = 28 \text{ m}^2$$

33. $1 \times 10^{17} \cancel{W} \times \dfrac{1 \text{kW}}{1000 \cancel{W}} = 1 \times 10^{14} \text{ kW}$

$1 \times 10^{14} \text{ kW} \times \dfrac{24 \text{ hours}}{1 \text{ day}} \times 365 \text{ days} = 8.76 \times 10^{17} \text{ kWh}$

$8.76 \times 10^{17} \text{ kWh} \times \dfrac{3.6 \times 10^{6} \text{ J}}{1 \text{ kWh}} \times \dfrac{1 \text{ Quad}}{1.06 \times 10^{18} \text{ J}} = 3 \times 10^{6} \text{ Quads each year}$

The actual energy consumption is very small compared to the amount of solar energy falling on the earth.

$\dfrac{325 \text{ Quads}}{3 \times 10^{6} \text{ Quads}} \times 100\% = 0.011\% \text{ consumption}$

35. $5.0 \, \cancel{m^2} \times \dfrac{1000 \cancel{W}}{\cancel{m^2}} \times \dfrac{1 \text{kW}}{1000 \cancel{W}} \times 0.50 \times \dfrac{8 \text{ hr}}{1 \text{ day}} \times \dfrac{30 \text{ day}}{1 \text{ month}} = 6.0 \times 10^{2} \text{ kWh / month}$

Yes, the hot tub receives enough energy.

37. $\dfrac{100 \cancel{W}}{0.15} \times \dfrac{1 \text{ m}^2}{1000 \cancel{W}} = 0.67 \text{ m}^2$

SOLUTIONS TO FEATURE PROBLEMS:

43. a) $49 \text{ quads} - 13 \text{ quads} = 36 \text{ quads}$

$\dfrac{36 \text{ quads}}{50 \text{ yr}} = 0.72 \text{ quads/year}$

b) $\dfrac{36 \text{ quads}}{13 \text{ quads}} = 280\% \text{ increase}$

$\dfrac{280\%}{50 \text{ yr}} = 5.6\% \text{ / year}$

REVIEW TESTS FOR CHAPTER 10

Match the following statements or phrases to the end-of-chapter key terms.

1. The term for the process in which silicon is mixed with small amounts of other elements.

2. The use of electric current to split a compound into its constituent elements.

3. The utilization of sunlight by plants to convert CO_2 and H_2O into energetic glucose molecules and O_2.

4. The type of nuclear reactor that converts non-fissile uranium-238 into fissile plutonium-239.

5. The term used when an n-type silicon sample is brought in contact with a p-type sample.

6. The type of semi-conductor that is electron-rich.

7. The type of semi-conductor that is electron-poor.

ANSWERS TO MATCHING

1. doping
2. electrolysis
3. photosynthesis
4. breeder reactor
5. p-n junction
6. n-type
7. p-type

SELF-TEST QUESTIONS

Completion: Write the word, phrase, or number in the blank that will complete the statement or answer the question.

1. Power out ÷ Power in = _____

2. Photovoltaic cells are composed of _____.

3. Write the balanced chemical equation for the process of photosynthesis.

4. In a p-n junction, there is a tendency for electrons to move from the _____ side to the _____ side.

5. About one-fourth of U.S. energy is used for _____.

6. A solar power tower has 50 heliostats each of which has an active area of 3.2 m². The tower produces 29.7 kW of power. Assuming that the solar power falling on the active area is 1000 W/m², what is the efficiency of the tower?

Multiple choice: Select the correct answer from the choices listed.

7. What is the main obstacle in using solar energy?

   a) its abundance
   b) its low concentration
   c) its high temperatures
   d) its toxic emissions

8. What is a disadvantage of using hydropower as an energy source?

   a) non-renewable
   b) inefficient
   c) atmospheric contamination
   d) threat to marine life

9. Which one of the following chemical equations correctly portrays the electrolysis of water?

   a) $2 H_2O + energy \rightarrow 2 H_2 + O_2$

   b) $2 H_2O \rightarrow 2 H_2 + O_2 + energy$

   c) $2 H_2 + O_2 \rightarrow 2 H_2O + energy$

   d) $2 H_2 + O_2 + energy \rightarrow 2 H_2O$

10. The burning of wood to produce heat is a form of _____ energy.

    a)   hydroelectric
    b)   geothermal
    c)   biomass
    d)   nuclear

11. The fermentation of corn or sugar cane by yeasts produces which of the following product(s)?

    a)   methanol and carbon dioxide
    b)   ethanol and carbon dioxide
    c)   ethanol only
    d)   carbon dioxide only

12. What is the largest single household energy use?

    a)   appliances
    b)   water heating
    c)   space heating
    d)   air conditioning

13. What is the function of a solar parabolic trough?

    a)   produce electricity
    b)   electrolyze water
    c)   store solar energy
    d)   track and focus sunlight

14. What is a disadvantage in using nuclear power as an energy source?

    a)   limited supply of U-235
    b)   significant carbon dioxide emissions
    c)   nonrenewable
    d)   very inefficient

15. Approximately how much of the sun's power does the earth intercept?

    a)   one part in a thousand
    b)   one part in a million
    c)   one part in a billion
    d)   one part in a trillion

ANSWERS TO SELF-TEST QUESTIONS

1. efficiency
2. semi-conductors
3. sunlight + 6 CO_2 + 6 H_2O → $C_6H_{12}O_6$ + 6 O_2
4. n-type, p-type
5. transportation

6. $3.2 \text{ m}^2 \times 50 = 160 \text{ m}^2$

$$160 \text{ m}^2 \times \frac{1000 \text{ W}}{\text{m}^2} = 1.60 \times 10^5 \text{ W}$$

$$\frac{2.97 \times 10^4 \text{ W}}{1.60 \times 10^5 \text{ W}} \times 100\% = 19\%$$

$$29.7 \text{ kW} \times \frac{1000 \text{ W}}{1 \text{ kW}} = 2.97 \times 10^4 \text{ W}$$

7. b
8. d
9. a
10. c
11. b
12. c
13. d
14. a
15. c

CHAPTER 11

THE AIR AROUND US

ANSWERS TO QUESTIONS:

1. Pressure is the net force produced by the constant pounding or striking of gas molecules. It is the direct result of constant collisions between gas molecules and the surfaces around them.

3. An increase in altitude causes a decrease in pressure, which is responsible for the pain in our ears. Gas molecules are constantly being trapped in our ears, usually at the same pressure as the gas external to our ears. Equal pressures normally result because the molecules colliding on either side of our eardrum is exactly the same. However, as the air pressure changes, there is an imbalance that results in more collisions on one side of the eardrum than the other. Thus, the eardrum is stressed, resulting in pain.

5. The height of the mercury column supported by the gases in the atmosphere varies with weather conditions and altitude. High-pressure areas direct storms away and are a sign of good weather, while low pressure areas draw storms in, resulting in rain. Also, winds are a direct result of changes in pressure form one geographical region to another. Air molecules migrate from regions of high pressure to regions of low pressure.

7. If the temperature of a gas is increased, its volume will increase in direct proportion. If an inflated balloon is placed above a heater, the balloon will expand as the temperature of the gas within the balloon increases. Afternoon winds in coastal regions are a direct result of Charles' Law. Inland air masses are warmed by the sun and therefore rise; air from the coast, which remains cool due to the high heat capacity of oceans, rushes in to fill the void, resulting in wind.

9. Hot air balloons achieve buoyancy based on Charles' Law. When the air in the balloon is heated, its volume increases resulting in a lower density and increased buoyancy. Afternoon winds in coastal regions are a direct result of Charles' Law. As inland air masses are warmed by the sun they expand and rise, then the air from the coast, which remains cool due to the high capacity of oceans, rushes in to fill the void, resulting in wind.

11. Nitrogen fixation is the process of transforming N_2 to other nitrogen – containing compounds. The main objective is to change nitrogen from the inert N_2 molecule to a form usable by plants and animals. The Haber process (production of ammonia) is

one example of nitrogen fixation. Nitrogen fixation also results from the high-temperature combustion process in automobile engines. There are also natural nitrogen fixation processes such as lightning and nitrogen-fixing bacteria.

13. Oxygen, part of our atmosphere, is responsible for the rusting of iron and the dulling of paint.

15. Argon, neon, and helium gases are all very unreactive or inert. Neon is employed in luminescent lighting (neon signs), and argon is used to provide the noncorrosive atmosphere for electronic systems and units. Helium is used in balloons and blimps and used as a coolant to achieve temperatures near zero Kelvin.

17. The troposphere is the region of the atmosphere in which nearly all of us live out our entire lives. The stratosphere is the region of the atmosphere where commercial jet airplanes fly and which contains ozone (O_3). Ozone absorbs harmful UV light. The mesosphere is the region where meteors burn up. High-speed ionic gas particles emit light in the ionosphere. The ionosphere is the boundary of outerspace.

19. a) Sulfur dioxide is a strong irritant that affects the respiratory/cardiovascular system. It is also the major precursor to acid rain.

    b) PM-10's adverse health effects include aggravation of existing respiratory and cardiovascular disease, alterations in the body's immune system, damage to lung tissue and cancer. PM-10 can also lower visibility and soil building materials (make them less pure).

    c) Carbon monoxide is very toxic and its adverse health effects include impairment of visual perception, decreased work capacity, decreased manual dexterity and reduced learning ability.

    d) Ozone can reduce lung function, sometimes causing permanent structural damage to the lungs after long-term exposure. Affects of ozone exposure include: chest pain, coughing, nausea and pulmonary congestion. Ozone also damages rubber, agricultural crops and many tree species.

    e) Nitrogen dioxide is also a lung and eye irritant. It is a precursor to ozone formation and contributes to acid rain.

    f) Lead accumulation in the body can damage the kidneys, liver and nervous system. High exposure to lead causes neurological damage producing seizures, mental retardation and other behavior disorders.

93

21. Over the past ten years Pb and CO pollution have been significantly reduced while SO_2, PM-10, O_3 and NO_2 pollutants have shown moderate reductions. PM-10, CO, Pb and especially O_3 are still above acceptable levels in certain cities.

23. "Photochemical smog" is a stagnant air mass produced by photochemical processes (degradation of compounds by sunlight). This condition occurs when sunlight reacts with nitrogen oxides and unburned hydrocarbons.

25. The following series of equations illustrate how CFC's deplete the ozone layer in the stratosphere:

$$CF_2Cl_2 \; + \; UV\text{-}light \rightarrow Cl + CF_2Cl$$
$$Cl + O_3 \rightarrow ClO + O_2$$
$$O_3 \; + \; UV\text{-}light \rightarrow O \; + O_2$$
$$ClO + O \rightarrow Cl + O_2$$

The net result is that one Cl atom destroys two ozone molecules and is regenerated in the process, allowing it (Cl) to react over and over again. Thus a chlorine atom is a catalyst in the complete cyclic process.

27. Ozone has been depleted over Antarctica due to its unique geography and climate. Unique clouds form, called polar stratospheric clouds (PSC), which allow for the catalytic conversion of repository chlorine to reactive chlorine atoms. Most other parts of the planet have higher temperatures, which prevent the formation of PSCs.

29. The phase-out of CFC's began in1978 with the banning of CFC's from aerosols such as deodorants and hairsprays. The Montreal protocol involved 23 nations agreeing to decrease CFC usage to 50% by the year 2000. The Clean Air Act Amendment of 1990 also contained legislation to enact this elimination. But, in 1992 President Bush pushed the deadline for complete phase out of CFC's to January 1, 1996.

31. HCFC's are hydrochlorofluorocarbons. They contain chlorine and threaten the ozone layer. However, HCFC's are more reactive because of the presence of hydrogen and tend to decompose in the troposphere.

33. The ozone layer will take years to recover because CFC's are still leaking out of older refrigerators and air conditioners and into the atmosphere.

35. Chlorine from volcanoes is primarily in the form of HCl. Hydrogen chloride is a very reactive and water-soluble compound. Also, hydrogen chloride is quickly washed out by rain and condensed steam from the eruption of the volcano itself. On the other

hand, CFC's are not water soluble, and are not removed from the atmosphere by rainfall.

SOLUTIONS TO PROBLEMS:

37. $11 \text{ atm} \times \dfrac{760 \text{ torr}}{1 \text{ atm}} = 8.4 \times 10^3 \text{ torr}$

$11 \text{ atm} \times \dfrac{14.8 \text{ psi}}{1 \text{ atm}} = 1.6 \times 10^2 \text{ psi}$

39. a) $0.95 \text{ atm} \times \dfrac{760 \text{ torr}}{1 \text{ atm}} = 7.2 \times 10^2 \text{ torr}$

b) $0.95 \text{ atm} \times \dfrac{760 \text{ mmHg}}{1 \text{ atm}} = 7.2 \times 10^2 \text{ mmHg}$

c) $0.95 \text{ atm} \times \dfrac{29.92 \text{ in Hg}}{1 \text{ atm}} = 28 \text{ in Hg}$

d) $0.95 \text{ atm} \times \dfrac{14.8 \text{ psi}}{1 \text{ atm}} = 14 \text{ psi}$

e) $0.95 \text{ atm} \times \dfrac{101,325 \text{ N/m}^2}{1 \text{ atm}} = 96000 \text{ N/m}^2 \text{ or } 9.6 \times 10^4 \text{ kPa}$

Note: $1 \text{ N/m}^2 = 1 \text{ Pa}$

41. Apply Boyle's Law,

$$V_2 = \dfrac{P_1 V_1}{P_2}$$

$$V_2 = \dfrac{1.2 \text{ atm} \times 22.4 \text{ L}}{1.7 \text{ atm}} = 16 \text{ L}$$

43. Apply Boyle's Law,

$$V_2 = \frac{P_1 V_1}{P_2}$$

$$V_2 = \frac{1\,atm \times 1.5\,L}{3.7\,atm} = 0.41\,L$$

45. Apply Charles' Law, and remember that all temperatures must be in Kelvin in (°C + 273 = K).

$$60 + 273 = 333\,K = T_1$$
$$20 + 273 = 293\,K = T_2$$
$$V_2 = \frac{V_1 T_2}{T_1} = \frac{64{,}263\,m^3 \times 293\,K}{333\,K} = 56{,}544\,m^3$$

47. Apply Charles' Law, and remember that all temperatures must be in Kelvin in (°C + 273 = K).

$$25 + 273 = 298\,K = T_1$$
$$45 + 273 = 318\,K = T_2$$
$$V_2 = \frac{V_1 T_2}{T_1} = \frac{2.8\,L \times 318\,K}{298\,K} = 3.0\,L$$

49. Apply the combined gas law,

$$T_1 = 20 + 273 = 293\,K$$
$$T_2 = 30 + 273 = 303\,K$$
$$V_2 = \frac{P_1 V_1 T_2}{T_1 P_2} = \frac{1.0\,atm \times 0.50\,L \times 303\,K}{293\,K \times 0.83\,atm} = 0.62\,L$$

SOLUTIONS TO FEATURE PROBLEMS:

57. At the higher temperature, the particles should be moving faster. Since the pressure is higher, the volume should be lower, resulting in a higher density of particles.

REVIEW TESTS FOR CHAPTER 11

Match the following statements or phrases to the end-of-chapter terms.

1. The net force created by the constant pelting of gas molecules.

2. A device used to measure the pressure of the atmosphere.

3. If the pressure of a gas is increased, its volume will decrease in direct proportion.

4. If the temperature of a gas is increased, its volume will increase in direct proportion.

5. The process whereby bacteria, present in soil, break nitrogen's triple bond to form nitrate compounds.

6. The lowest part of the atmosphere, closest to the earth.

7. The part of the atmosphere above the stratosphere, at 50-80 km.

8. The part of the atmosphere that borders on outer space.

9. The region of the atmosphere containing the protective ozone layer.

ANSWERS TO MATCHING

1. pressure
2. barometer
3. Boyle's Law
4. Charles' Law
5. nitrogen fixation
6. troposphere
7. mesosphere
8. ionosphere
9. stratosphere

SELF-TEST QUESTIONS

Completion: Write the word, phrase, or number in the blank that will complete the statement or answer the question.

1.   The basic units of pressure are those of _____ per _____.

2.   Convert 800 mm Hg to atm.

3.   Write the mathematical statement for Charles' Law.

4.   What are the four fundamental properties associated with a sample of gas?

5.   What is the name of the federal law that regulates atmospheric emissions?

6.   A balloon is inflated to a volume of 2.5 liters in an airplane where the pressure is 0.70 atm. What will be the volume, in liters, of the balloon upon returning to sea level where the pressure is 1.0 atm?

Multiple choice: Select the correct answer from the choices listed.

7.   Which one of the following units is <u>not</u> a unit of pressure?

    a)   cm Hg
    b) N/m^2
    c)   lb
    d)   Pa

8.   What is the most abundant gas in our atmosphere?

    a)   oxygen
    b)   nitrogen
    c)   carbon dioxide
    d)   argon

9.   Which atmospheric gas absorbs ultraviolet radiation?

    a)   oxygen
    b)   carbon dioxide
    c)   nitrogen
    d)   ozone

10.   All weather phenomena occur in what section of the atmosphere?

a)   troposphere
b)   stratosphere
c)   mesosphere
d)   ionosphere

11.   What is the chemical formula for Freon-12?

a) CH_2FCF_3
b) CCl_2F_2
c) CH_2Cl_2
d) CH_2F_2

12.   Which one of the following is a typical property of CFCs?

a)   pungent odor
b)   toxic
c)   exhibit various colors
d)   chemically inert

13.   As altitude increases, pressure _____.

a)   increases
b)   decreases
c)   remains the same

14.   Select the mathematical statement for the combined gas law.

a) $P_1V_1T_1 = P_2V_2T_2$
b) $P_1V_1T_2 = P_2V_2T_1$
c) $P_2V_1T_1 = P_1V_2T_2$
d) $P_1V_1 = P_2V_2$

15.   What chemical compound has replaced CFCs in the refrigeration industry?

a) CF_3CF_3
b) CH_3CH_3
c) CH_2FCH_2F
d) CH_2FCF_3

16. Only chlorine that makes it to the _____ depletes ozone.

    a) troposphere
    b) stratosphere
    c) mesosphere
    d) ionosphere

17. What volume will 3.00 L of a gas at 30°C occupy if the temperature is increased to 90°C?

    a) 1.00 L
    b) 9.00 L
    c) 6.00 L
    d) 3.59 L

18. If the pressure of a gas is 76.7 mm Hg, what is the pressure expressed in torr?

    a) 7.67 torr
    b) 76.7 torr
    c) 767 torr
    d) 7670 torr

ANSWERS TO SELF-TEST QUESTIONS

1. force, unit area

2.
$$800 \text{ mm Hg} \times \frac{1 \text{ atm}}{760 \text{ mm Hg}} = 1.05 \text{ atm}$$

3.
$$\frac{V_1}{T_1} = \frac{V_2}{T_2}$$

4. Amount of a gas, volume, temperature, pressure

5. The Clean Air Act

6.
$$V_2 = \frac{P_1V_1}{P_2}$$

$$V_2 = \frac{0.70 \text{ atm} \times 2.5 \text{ L}}{1.0 \text{ atm}} = 1.8 \text{ L}$$

7. c
8. b
9. d
10. a
11. b
12. d
13. b
14. b
15. d
16. b
17. d
18. b

CHAPTER 12

THE LIQUIDS AND SOLIDS AROUND US: ESPECIALLY WATER

ANSWERS TO QUESTIONS:

1. Liquids are spherical in shape due to cohesive forces. Molecules squeeze together into spheres to maximize their contact. A sphere has the lowest surface-area-to-volume ratio and, therefore, allows the maximum number of molecules to be completely surrounded by other molecules.

3. The differences between solids, liquids, and gases, from a molecular viewpoint, lies in the interaction of the molecules. The interaction between molecules is dependent not only on the kind of molecule present, but also on their distance of separation. For gases, the attractions between molecules are weak since large distances separate the molecules. This molecular attraction becomes progressively stronger going to a liquid and then to a solid, as the molecules become closer and closer together.

5. a) Boiling point is the temperature at which a liquid boils. This boiling occurs when molecules are able to completely overcome their cohesive forces and leave the liquid state.

    b) Melting point is the temperature at which a solid melts. It depends on the strength of the cohesive forces of the atoms or molecules that compose it.

    c) Cohesive force is a force of attraction that exists between atoms or molecules. The strength of this force is due to the kind of atoms or molecules present and their distance of separation.

7. a) Dispersion forces are the result of small fluctuations in the electron clouds of atoms and molecules. These fluctuations result in uneven distribution of electrons in an atom or molecule, which results in an instantaneous dipole.

    b) Dipole forces are a result of atoms of different electronegativities. In this case polar bonds are formed in which electrons are unevenly shared between the bonding atoms. This results in partial negative and partial positive charges on the two bonding atoms. However, the presence of polar bonds within a molecule may or may not make the entire molecule polar, depending on molecular structure.

    c) Hydrogen bonding is a cohesive attraction between molecules and is not a chemical bond that holds atoms together to form molecules. A hydrogen bond only exists in polar molecules which contain hydrogen atoms bonded to either F, O, or N atoms.

The strengths of the cohesive forces are as follows:

dispersion forces < dipole forces < hydrogen bonding

9. If you wish to dissolve stains by a non-polar compound such as grease you must use either a non-polar solvent or a soap or detergent. One end of the soap molecule is highly polar and dissolves in the water, whereas the other part of the molecule is a long, non-polar, hydrocarbon chain that dissolves in the grease. The grease is then emulsified and swept away by rinsing.

11. The sweat glands are involved in the cooling process of the human body. The sweat glands secrete water, which evaporates from the skin, absorbing heat and cooling the body.

13. Water is truly an unusual molecule in that for such a low molecular weight compound, it exists as a liquid at room temperature and has an anonymously high boiling point. These properties can be explained by the fact that water contains polar bonds and can hydrogen bond to other water molecules. In addition, while almost every other liquid contracts upon freezing, water expands. Consequently, ice has a lower density than liquid water. Another unique property of water is its ability to dissolve many organic and inorganic compounds. Many nutrients and biological molecules, essential for human body survival, dissolve in water. Also, if ice sank in water, entire lakes could freeze solid and eradicate marine life. The slow and constant erosion of mountains is due in part to the expansion of water upon freezing. The expansion of ice is also responsible for the damage to cells in biological tissue when frozen.

15. Our water supply is a reusable resource that is constantly being redistributed over earth. The hydrologic cycle is the movement of moisture from large reservoirs of water, such as oceans and seas, to the higher-elevated inland regions. Following a cyclic pattern, water then flows back to the sea, eroding as it goes.

17. <u>Classification</u>   <u>Hardness (ppm $CaCO_3$)</u>

Classification	Hardness (ppm $CaCO_3$)
Very soft	< 15
Soft	15 – 50
Medium	50 – 100
Hard	100 – 200
Very hard	> 200

19. | Contaminants | Examples |
| --- | --- |
| Biological | Giardia, Legionella |
| Inorganic | Mercury, Lead |
| Organic | Volatile – Benzene |
| | Non-volatile – chlorobenzene |
| Radioactive | Uranium, Radon |

21. Most micro-organisms (biological contaminants) are killed by boiling water.

23. The Safe Drinking Water Act was enacted in 1974 by Congress to ensure high water quality. The SDWA authorized the EPA to establish Maximum Contaminant Levels (MCL) for specific contaminants likely to be found in public drinking water systems. Under the SDWA, public water suppliers must routinely sample and test the water supplied to your tap. If any contaminant is above the MCL, the supplier must act to correct the problem as soon as possible.

25. If you receive your water from a public water provider, the EPA believes home water treatment for health reasons is not necessary. However, it may be important to perform home water treatment to reduce hardness, improve clarity, or to eliminate an undesirable taste in the water.

27. Environmental groups, such as the EWA and NRDC, believe that water quality is not as good as it could be, and consequently push for more regulation of businesses and industry. These groups issued reports stating that tap water contains toxic contaminants and bacteria that cause death and many cases of illness in the U.S. population.

SOLUTIONS TO PROBLEMS:

29. Bromine (Br_2) is a liquid at room temperature. It has the highest molecular weight among O_2, F_2 and Cl_2, and will have the strongest dispersion forces. This results in stronger cohesive forces, keeping bromine molecules in the liquid state.

31. Water (H_2O) would have the highest boiling point. Its anomalously high boiling point is due to substantial intramolecular hydrogen bonding.

33. Volatility is a function of the strength of the cohesive forces between molecules. These liquids are all hydrocarbons. There are no functional groups present so one can use molecular weight to decide upon volatility. Decane has the highest molecular weight and thus the strongest dispersion forces, resulting in it being the least volatile of the liquids.

35. a) CO_2 will be non-polar because the C-O bonds cancel each other out.

b) CH_2Cl_2 will be polar because of the chlorine.

c) Octane is a hydrocarbon therefore it is non-polar.

d) CH_3CHF_2 will be polar because of the fluorine.

37. The non-polar substance with the lower molecular weight will be smelled first due to the fact that it will be more volatile. After several hours the substance with the higher molecule weight will been smelled because the lighter substance will have totally evaporated.

39. $$\text{Molarity} = \frac{\text{moles of solute}}{\text{liter of solution}}$$

$$34 \text{ g NaNO}_3 \times \frac{1 \text{ mole NaNO}_3}{85 \text{ g NaNO}_3} = 0.40 \text{ moles NaNO}_3$$

$$\frac{0.40 \text{ moles NaNO}_3}{2.8 \text{ L}} = 0.14 \text{ M NaNO}_3$$

41. $$\frac{1.3 \text{ mole glucose}}{1 \text{ L}} \times 7.8 \text{ L solution} = 10.14 \text{ moles glucose}$$

$$10.14 \text{ moles glucose} \times \frac{180 \text{ g glucose}}{1 \text{ mole glucose}} = 1.8 \times 10^3 \text{ g glucose}$$

43. $$3.8 \text{ L} \times \frac{1000 \text{ mL}}{1 \text{ L}} \times \frac{1.0 \text{ g}}{1 \text{ mL}} = 3.8 \times 10^3 \text{ g solution}$$

$$3.8 \times 10^3 \text{ g solution} \times \frac{.011 \text{ g NaF}}{1 \text{ g solution}} = 42 \text{ g NaF}$$

45. $$\frac{1.1 \times 10^{-3}\text{ g}}{15\text{ g}} \times \left(10^{6}\right) = 73\text{ ppm}$$

This sample would receive a medium classification.

47. $$1.5\text{ L} \times \frac{1000\text{ mL}}{1\text{ L}} \times \frac{1\text{ g}}{1\text{ mL}} = 1500\text{ g solution}$$

$$1500\text{ g solution} \times \frac{245\text{ g Na}}{10^{6}\text{ g solution}} \times \frac{1000\text{ mg}}{1\text{ gram}} = 368\text{ mg Na}$$

49. $$\frac{0.012\text{ mg}}{.225\text{ L}} = \frac{0.053\text{ mg}}{\text{L}}$$

No, the water is not safe for human consumption. The TCE concentration is too high.

51. No, boiling the water would actually increase the nitrate concentration.

SOLUTION TO FEATURE PROBLEMS:

59. The boiling point of lead is extremely higher than water. Therefore when the solution is boiled only water will become gas. When half the water is boiled away there will be 6 remaining water molecules and all three lead molecules will still be present. Consequently the lead concentration is increased.

REVIEW TESTS FOR CHAPTER 12

Match the following statements or phrases to the end-of-chapter key terms.

1.   Type of attractive force between molecules.

2.   The repeating pattern of molecules or atoms within solids.

3.   Type of solid with a well-ordered arrangement of molecules or atoms.

4.   Process in which molecules at the surface of a liquid overcome attractive forces and shoot off into the air.

5.   The temperature at which a solid melts.

6.   The weakest type of cohesive force present between all atoms and molecules.

7.   Small fluctuations in the electron cloud of atoms and molecules that result in instances in which electrons are not evenly distributed in the molecule or atom.

8.   Cohesive attraction between molecules which contain H atoms bonded to either F, O, or N atoms.

9.   The term applied to liquids that do not vaporize easily and have a low vapor pressure.

10.   A mixture in which a solid dissolves in a liquid, or one liquid dissolves in another.

11.   The majority component of a solution.

12.   The amount of solute relative to the amount of solvent in a solution.

13.   The number of moles of solute divided by the number of liters of solution.

14.   The process in which food is frozen very quickly.

15.   The cycling of water between the atmosphere, land, and oceans on the earth.

16.   Water containing appreciable amounts of calcium and magnesium ions.

17.   An ion-exchange material used for water softening.

18.   A naturally-occurring process in which water flows from an area of low solute concentration to high solute concentration.

ANSWERS TO MATCHING

1.   cohesive
2.   crystalline structure
3.   crystalline solid
4.   evaporation
5.   melting point

6. dispersion force
7. instantaneous dipole
8. non-volatile
9. solution
10. solvent
12. concentration
13. molarity
14. flash freezing
15. hydrologic cycle
16. hard water
17. zeolite
18. osmosis

SELF-TEST QUESTIONS

Completion: Write the word, phrase, or number in the blank which will complete the statement or answer the question.

1. Substances with constituent molecules or atoms having weak cohesive forces, have _____ melting and boiling points.

2. Which has the higher boiling point, Kr or Xe?

3. Classify the molecules as polar or non-polar.

   a) I_2
   b) CH_3Br

4. Which of the following molecules would you expect to have the highest boiling point?

   $CH_3CH_2CH_3$ or $CH_3CH_2NH_2$

5. Which of the following molecules would be most volatile?

   $CH_3CH_2CH_2CH_3$ or $CH_3CH_2CH_2CH_2OH$

6. A solution of sodium hydroxide is prepared by mixing 2.00 g of NaOH with 10.0 g of water. What is the percent by mass concentration of NaOH?

7. What is the molarity of a solution prepared by mixing 45 grams of sugar (M.W. = 342) with enough water to make 350 mL of solution?

8.   The biological contaminants that can find their way into U. S. water include bacteria such _____ or _____.

9.   Which one of the following molecules is a gas at room temperature?

      CH_3OCH_3 $CH_3CH_2OCH_2CH_3$

10.   What is the purpose of zeolite in the softening of water?

Multiple choice: Select the correct answer from the choices listed)

11.   Which one of the following organic contaminants would be classified as non-volatile?

      a)   benzene
      b)   dichloromethane
      c)   ethylbenzene
      d)   carbon tetrachloride

12.   The minority component in a solution is generally called the

      a)   solute
      b)   solvent
      c)   salt
      d)   molar concentrate

13.   How many moles of NaOH does 2.0 L of a 3.0 M solution of NaOH contain?

      a)   1.0
      b)   2.0
      c)   3.0
      d)   6.0

14.   Which one of the following molecules would exhibit hydrogen bonding?

      a) H_2S
      b) CH_3CH_2Cl
      c) CH_3OCH_3
      d) CH_3NH_2

15. Which one of the following statements concerning the properties of water is <u>false</u>?

    a)   Water dissolves both organic and inorganic compounds.
    b)   Water contracts upon freezing.
    c)   Water has a higher boiling point than H_2S.
    d)   Water is a polar molecule.

16. What is the Maximum Contaminant Level (MCL) in mg/L for lead, in drinking water?

    a)   0
    b)   0.005
    c)   0.1
    d)   10

17. Which one of the following molecules would have the largest dispersion force?

    a)   Ar
    b)   Kr
    c)   Xe
    d)   Rn

18. Select the non-polar molecule from the following list.

    a) Br_2
    b)   CO
    c) CO_2
    d) $CH_3CH_2NH_2$

19. Which one of the following water contaminants poses an immediate health risk?

    a)   lead
    b)   viruses
    c)   radon
    d)   asbestos

20. Hard water is classified by the amount, in ppm, of what compound?

    a) $CaCO_3$
    b) $CaCl_2$
    c) $MgCO_3$
    d)   NaCl

ANSWERS TO SELF-TEST QUESTIONS

1. low
2. Xe
3. a) non-polar   b)   polar
4. $CH_3CH_2NH_2$
5. $CH_3CH_2CH_2CH_3$

6.
$$\frac{2.00\ g}{2.00\ g + 10.0\ g} \times 100\% = 16.7\%$$

7.
$$\frac{45\ g}{342\ g/mol} = 0.13\ moles$$

$$350\ mL \times \frac{1\ L}{1000\ mL} = 0.350\ L$$

$$Molarity = \frac{0.13\ moles}{0.350\ L} = 0.37\ moles/liter\ or\ 0.37\ M$$

8. *Giardia, Legionella*
9. CH_3OCH_3
10. removes calcium and magnesium ions through an exchange process with sodium ions.
11. c
12. a
13. d
14. d
15. b
16. a
17. d
18. c
19. b
20. a

CHAPTER 13

ACIDS AND BASES: THE MOLECULES RESPONSIBLE FOR SOUR AND BITTER

ANSWERS TO QUESTIONS:

1.   The more acidic the food, the more sour the taste.

3.   The following list includes the common properties of bases:

   a)   They have a slippery feel.
   b)   They have a bitter taste
   c)   They have the ability to react with acids
   a)   They have the ability to turn litmus paper blue.

5.   <u>Acid</u>   <u>Use</u>

   Hydrochloric Acid   Cleaning metals; preparation of foods;
                                    refining of ores

   Sulfuric Acid   Manufacture of fertilizers, explosives, dyes
                                    and glue

   Nitric Acid   Manufacture of fertilizers, explosives and
                                    dyes

   Phosphoric Acid   Manufacture of fertilizer and detergents;
                                    flavor additive for food and drinks.

   Acetic Acid   Present in vinegar, manufacture of plastics
                                    and rubber, preservative in foods, solvent for
                                    resins and oils.

7.   <u>Base</u>   <u>Use</u>

   Sodium Hydroxide   Neutralization of acids; petroleum
                                    processing; manufacture of soap and
                                    plastics.

   Potassium Hydroxide   Manufacture of soap; cotton processing;
                                    electroplating; paint remover.

| Sodium Bicarbonate | Antacid; source of CO_2 in fire extinguishers and cleaning products. |
| Ammonia | Detergent; removing of stains; extracting plant color; manufacture of fertilizers, explosives and synthetic fibers. |

9. A Bronsted-Lowry Acid is a proton donor; a Bronsted-Lowry base is a proton acceptor.

11. For every change of one unit of the pH scale, the $[H_3O^+]$ concentration changes by a factor of ten. Thus a solution with a pH of 4 has $[H_3O^+] = 1 \times 10^{-4}$ M and a solution with a pH of 3 is ten times more acidic with a $[H_3O^+] = 1 \times 10^{-3}$M.

13. Citric acid is responsible for the sour taste of lemons, limes and oranges.

15. Acetic acid can be found in vinegar and salad dressing.

17. Hydrochloric acid (HCl) is found in high concentrations in the stomach. Phosphoric acid is often added to soft drinks and beer to impart tartness. Carbonic acid is present in all carbonated beverages.

19. Alkaloids are nitrogen containing organic compounds, which are basic and often poisonous.

21. Leavening agents produce pockets of carbon dioxide gas in the dough of baked goods. A common leavening agent is baking powder, which contains $NaHCO_3$, NaAl $(SO_4)_2$, and $CaHPO_4$. Both NaAl $(SO_4)_2$, and $CaHPO_4$ function as acids when combined with water. The H_3O^+ then reacts with the HCO_3^- from the sodium bicarbonate to produce CO_2 and H_2O. This reaction, in conjunction with rising temperatures, causes the carbon dioxide gas to expand, allowing the baked goods to be quite larger than the original dough.

23. Rain is naturally slightly acidic due to the presence of carbon dioxide in the atmosphere. This CO_2 is able to combine with water to form carbonic acid. The pH of water saturated with CO_2 is approximately 5.6.

25. Many soils and natural waters contain significant amounts of basic ions (e.g. HCO_3^-) that come from the weathering of rocks. Thus the lake or soil can neutralize much of the incoming acid and minimize pollution problems.

27. The Congress has passed several amendments to the Clean Air Act to reduce acid rain damage in the U.S. These amendments include having electric utility companies cutting their sulfur dioxide emissions by one-half of the 1980 levels by the 2010.

SOLUTION TO PROBLEMS:

29. $H_2SO_4 + Ca(OH)_2 \rightarrow 2H_2O + CaSO_4$

$2H^+ + 2OH^- \rightarrow 2H_2O$

31. a) $C_5H_5N + H_2O \longrightarrow OH^- + C_5H_5NH^+$
base acid

b) $HC_2H_3O_2 + H_2O \longrightarrow H_3O^+ + C_5H_3O_2^-$
acid base

c) $KOH + HBr \longrightarrow H_2O + KBr$
base acid

33.

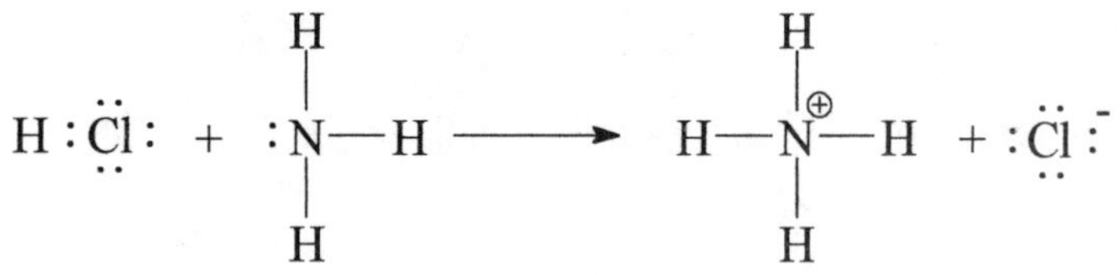

Acid Base

35. The pH of a 0.001M NH_3 solution would be less than 11, because a large fraction of NH_3 molecules do not react with water. Consequently, significant amounts of NH_3, NH_4^+, and OH^- are present in a solution of NH_3. The reaction does not go to completion; and a lower concentration of OH^- is present in solution. This means a decrease in the pH and a less basic solution.

37. a) pH = 3 acidic
b) pH = 11 basic
c) pH = 8 basic
d) pH = 1 acidic

39. a) $CaCO_3 + 2\ HCl \rightarrow CaCl_2 + H_2O + CO_2$

b) $Al(OH)_3 + 3\ HCl \rightarrow AlCl_3 + 3\ H_2O$

41. $4 NO_2 + O_2 + 2 H_2O \rightarrow 4 HNO_3$

SOLUTIONS TO FEATURE PROBLEMS:

47. HF is a weak acid. Very little of the dissolved solute has ionized.

REVIEW TESTS TO CHAPTER 13

Match the following statements or phrases to the end-of-chapter key terms.

1.   The reaction of an acid and base to form water and a salt.

2.   A particular type of paper that contains a dye which turns red in acid and blue in base.

3.   A family of organic compounds that contain nitrogen and show physiological activity.

4.   A substance that produces hydroxide ions (OH⁻) in solution.

5.   An acid defined as a proton donor.

6.   Acids that completely dissociate in water.

7.   The scale developed to express acidity or basicity in a compact way.

8.   Devices which spray a mixture of water and limestone into exhaust, trapping sulfur dioxide.

9.   A base defined as a proton acceptor.

10.   A base that completely dissociates in solution.

11.   The type of solution that has an $[H_3O^+] > 1 \times 10^{-7}$ M.

12.   The type of solution that has an $[H_3O^+] < 1 \times 10^{-7}$ M.

ANSWERS TO MATCHING

1. neutralization
2. litmus paper
3. alkaloids
4. Arrhenius base
5. Bronsted-Lowry acid
6. strong acids
7. pH scale
8. flue gas scrubbers
9. Bronsted-Lowry base
10. strong base
11. acidic
12. basic

SELF-TEST QUESTIONS

Completion: Write the word, phrase, or number in the blank which will complete the statement or answer the question.

1. Acids typically have a characteristic _____ taste.

2. Write the chemical formula for phosphoric acid)

3. Identify the Bronsted-Lowry acid and base in the following reaction.

$$HBr + H_2O \rightarrow Br^- + H_3O^+$$

4. Write the formula for barium hydroxide.

5. A neutral solution has a pH of _____.

6. Give the pH that corresponds to the following solution and classify it as acidic, basic, or neutral.

$$[H_3O^+] = 10^{-6}$$

7. Write a chemical reaction to show how $MgCO_3$ would neutralize HCl solution.

8. Write a chemical equation using Lewis structures for the reaction of water with formic acid)

116

9.   Write a chemical reaction to show how CO_2 forms carbonic acid in the atmosphere.

Multiple choice: Select the correct answer from the choices listed.

11.   Select the incorrect property of a base from the following list.

a)   turns litmus red
b)   have a slippery feel
c)   taste sweet
d)   react with acids to form water and salt

12.   Which one of the following compounds is a weak acid in water solution?

a) C_6H_6COOH
b) H_2SO_4
c) NH_3
d)   KOH

13.   Consider four solutions with the following pH values. Which solution has the lowest H_3O^+ concentration?

a)   3.0
b)   7.0
c)   9.0
d)   11.0

14.   Which one of the following compounds could function as either a Bronsted-Lowry acid or base in water solution?

a) H_2O
b)   HCl
c)   KOH
d) CH_3COOH

15.   Which substance listed would not be an active ingredient in antacids?

a) $NaHCO_3$
b) Na_2CO_3
c) $Mg(OH)_2$
d)   KCl

117

16. Which one of the following compounds contributes to acid rain?

    a) NO_2
    b)   HF
    c)   HCl
    d) CCl_4

17. Which one of the following would <u>not</u> function as an Arrhenius acid in water solution?

    a)   HI
    b)   HCOOH
    c) CH_3NH_2
    d) HNO_3

18. The acidity of a solution is normally specified by the concentration of H_3O^+ in what kind of units?

    a)   parts per million
    b)   mg/liter
    c)   moles/liter
    d)   grams/liter

19. What does the following chemical equation indicate about the relative acidity or basicity of HNO_2?

    $$HNO_2 + H_2O \Leftrightarrow NO_2^- + H_3O^+$$

    a) HNO_2 is a weak acid)
    b) HNO_2 is a strong acid)
    c) HNO_2 is a weak base.
    d) HNO_2 is a strong base.

20. What is the name of the acid present in soft drinks?

    a)   hydrochloric acid
    b)   phosphoric acid
    c)   lactic acid
    d)   salicylic acid

ANSWERS TO SELF-TEST QUESTIONS

1. sour
2. H_3PO_4
3. HBr – acid, H_2O – base
4. $Ba(OH)_2$
5. 7
6. pH = 6
7. $MgCO_3 + 2\,HCl \rightarrow MgCl_2 + CO_2 + H_2O$

8.

9. $CO_2 + H_2O \rightarrow H_2CO_3$
10. antacids, cleaning agents
11. c
12. a
13. d
14. a
15. d
16. a
17. c
18. c
19. a
20. b

CHAPTER 14

OXIDATION AND REDUCTION

ANSWERS TO QUESTIONS:

1. In a piece of iron, rusting occurs as electrons transfer from iron atoms to oxygen atoms. The oxidized iron atoms (iron atoms that lost electrons) bond with the reduced oxygen atoms (oxygen atoms that gained electrons) to form iron oxide or rust. Thus, rusting requires the transfer or conduction of electrons.

3. Some common processes that involve redox reactions are the burning of coal, the function of batteries, metabolism of foods, and the corrosion of metals.

5. Reduction is defined as the loss of oxygen, the gain of hydrogen or the gain of electrons (the most fundamental definition).

7. Carbon shares its four valence electrons with two sulfur atoms and in this sense the electrons are not completely a part of the carbon atom anymore. The carbon has partially given away its electrons; therefore, the carbon atom takes on a partial positive charge. Because the carbon atom partially looses some of its electrons, it is oxidized.

9. The oxidizing agent is the substance that gains electrons and is reduced. It oxidizes the other compound. The reducing agent is the substance that loses electrons and is oxidized. It reduces the other compound.

11. Respiration:
$$C_6H_{12}O_6 + 6O_2 \rightarrow 6CO_2 + 6H_2O$$

      oxidized   reduced

   Photosynthesis:
$$6CO_2 + 6H_2O \rightarrow C_6H_{12}O_6 + 6O_2$$

      reduced   oxidized

   These redox reactions illustrate the fact that animals and plants depend on each other for life. Animals use oxygen to oxidize carbon and plants use the water to reduce the carbon back again completing the cycle.

13. An automobile battery consists of lead metal at the anode, lead (II) oxide at the cathode, and dilute sulfuric acid. Because the dissociation of H_2SO_4 does not go to completion, both HSO_4^- (aq) and SO_4^{2-} (aq) are present, but typically the two half-reactions and net reaction are expressed in terms of SO_4^{2-} (aq) as shown below:

Oxidation (anode): $\qquad$ $Pb(s) + SO_4^{2-} \rightarrow PbSO_4 + 2e^-$

Reduction (cathode): $\qquad$ $PbO_2 + 4H^+ + SO_4^{2-} + 2e^- \rightarrow PbSO_4 + 2H_2O$

Net: $\qquad$ $Pb + PbO_2 + 4H^+ + 2SO_4^{2-} \rightarrow 2PbSO_4 + 2H_2O$

Electrons flow from the lead plates to the lead oxide plates. Lead (Pb) is oxidized and forms lead (II) ions (Pb^{2+}), which react, with sulfate (SO_4^{2-}) ions in solution to form solid lead (II) sulfate ($PbSO_4$). As the lead (II) oxide (PbO_2) is reduced, it also reacts with SO_4^{2-} ions to form solid $PbSO_4$.

15. Fuel cells are batteries (actually a better term is an energy converter) with reactants that are continually renewed. In the normal chemical process within a battery, the reactants are stored within and become depleted over time. However, in a fuel cell, the reactants are constantly supplied as needed to produce electricity while the products constantly flow out of the cell.

17. In a $H_2 - O_2$ fuel cell, the reaction of H_2 (g) and O_2 (g) in a basic medium produces $H_2O(\ell)$ according to the following redox reactions:

Oxidation: $2 H_2 + 4 OH^- \rightarrow 4 H_2O + 4e^-$

Reduction: $O_2 + 2 H_2O + 4e^- \rightarrow 4 OH^-$

Net: $\qquad$ $2H_2 + O_2 \rightarrow 2H_2O$

19. The overall reaction for the oxidation or rusting of iron is:

$$2Fe + O_2 + 2H_2O \rightarrow 2Fe(OH)_2$$

Eventually, $Fe(OH)_2$ forms Fe_2O_3, iron (III) oxide which we call rust.
It is clear from the overall reaction that the formation of rust requires water, thus keeping iron dry helps prevent rusting.

21. The theory concerning the relationship of aging and oxidation proposes that free radicals extract electrons from large molecules within cell membranes, oxidizing the molecules and making them reactive. When the large molecules react with each other the cell wall properties change and the result is a weakened and vulnerable body. Free radicals are atoms or molecules with unpaired electrons that easily oxidize other molecules by extracting electrons from them. The free radicals are produced from the combustion of oxygen with pollutants or toxins present in food, water and air.

121

SOLUTIONS TO PROBLEMS:

23. All of the processes listed involve the transfer of electrons from one substance to another and are classified as redox reactions.

25. a) $4 \text{ Al} + 3 \overset{..}{\text{O}}{=}\overset{..}{\text{O}} \longrightarrow 2 [\text{Al}^{3+}]_2 [\overset{..}{\underset{..}{\text{O}}}{}^{2-}]_3$
        oxidized reduced

    b) $2 \text{ Li} + \overset{..}{\underset{..}{\text{F}}}{-}\overset{..}{\underset{..}{\text{F}}} \longrightarrow 2 [\text{Li}^+] [\overset{..}{\underset{..}{\text{F}}}{}^-]$
        oxidized reduced

    c) $2 \text{ H}{-}\overset{\overset{\displaystyle H}{|}}{\underset{\underset{\displaystyle H}{|}}{C}}{-}\overset{\overset{\displaystyle H}{|}}{\underset{\underset{\displaystyle H}{|}}{C}}{-}\text{H} + 7 \overset{..}{\text{O}}{=}\overset{..}{\text{O}} \longrightarrow 4 \overset{..}{\text{O}}{=}\text{C}{=}\overset{..}{\text{O}} + 6 \, {}_{H}{\overset{\overset{..}{\text{O}}}{\diagdown}}{}_{H}$
            oxidized   reduced

27. a) H_2 – oxidized, W – reduced
    b) Na – oxidized, Cl_2 – reduced
    c) C – oxidized, Fe – reduced

29. a) Sn – oxidized, H^+ - reduced
    b) I^- - oxidized, Br_2 – reduced
    c) Mg – oxidized, Cu^{2+} - reduced

31. a) NH_3 – reducing agent, O_2 – oxidizing agent
    b) C_6H_5OH – reducing agent, O_3 – oxidizing agent

33. $C_6H_{12}O_6 + 6 O_2 \rightarrow 6 CO_2 + 6 H_2O$

    O_2 is the oxidizing agent because the oxygen is reduced. $C_6H_{12}O_6$ is the reducing agent because the carbon is oxidized.

35. Antioxidants are reducing agents, which means they get oxidized. I_2 is an oxidizing agent that gets reduced and will not work as an antioxidant.

37. The zinc cannot be replaced with plastic because the battery needs the zinc to act as the reducing agent to produce the electricity.

SOLUTIONS TO FEATURE PROBLEMS:

43. a) anode
    b) cathode
    c) Zn^{2+} ions
    d) 2 e-
    e) $\longrightarrow$

REVIEW TESTS FOR CHAPTER 14

Match the following statements or phrases to the end-of-chapter key terms.

1.   In this type of chemical reaction, electrons are transferred from one substance to another.

2.   The process in which oxygen is used to oxidize sugars for energy in our bodies.

3.   Oxidizing agents used to sterilize and sanitize materials.

4.   The gain of electrons by one substance.

5.   The term used to describe substances which tend to gain electrons easily.

6.   The gaining of hydrogen by one substance.

7.   Industrial process used to reduce atmospheric nitrogen into ammonia.

8.   The process in which plants reduce carbon to form glucose and oxygen.

9.   A general term applied to a type of battery used to produce an electrical current from a redox reaction.

10.   The electrode where reduction occurs.

11.   Atoms or molecules with unpaired electrons.

12.   The coating of iron with zinc whose oxide is structurally stable and imparts corrosion resistance.

ANSWERS TO MATCHING

1. oxidation-reduction or redox
2. respiration
3. disinfectants
4. reduction or reduced
5. oxidizing agents
6. reduction
7. Haber Process
8. photosynthesis
9. electrochemical cell
10. cathode
11. free radicals
12. galvanization

SELF-TEST QUESTIONS

Completion: Write the word, phrase, or number in the blank which will complete the statement or answer the question.

1. Identify the reducing agent in the following chemical reaction.

$$Fe_2O_3 + 3\ CO \rightarrow 2\ Fe + 3\ CO_2$$

2. Draw a Lewis structure for the following chemical reaction.

$$Mg + Cl_2 \rightarrow MgCl_2$$

3. For the following reaction, identify which element is being oxidized and which one is being reduced.

$$4\ Al + 3\ MnO_2 \rightarrow 3\ Mn + 2\ Al_2O_3$$

4. All reactions in which an element or compound combines with oxygen are examples of _____ reactions.

5. Write the balanced chemical equation for the Haber Process.

6. In an electrochemical cell, electrons always flow from the _____ (sign) to the _____ (sign).

7. Write the balanced reaction which occurs at the anode in a hydrogen-oxygen fuel cell.

8. Write the overall balanced chemical reaction for the rusting of iron.

Multiple choice: (Select the current answer from the choices listed.)

9. A common disinfectant would include which one of the following substances?

    a) H_2O_2 c) Cl_2

    b) I_2 d) $(C_6H_5CO)_2O_2$

10. For the following reaction, what substance is being oxidized?

$$Cl_2 + NaI \rightarrow I_2 + NaCl$$

    a) I^-
    b) I_2
    c) Cl^-
    d) Cl_2

11. Which one of the following processes does <u>not</u> represent a redox reaction?

    a)   combustion of natural gas
    b)   burning of coal
    c)   reaction between hydrogen and oxygen
    d)   dissolving sodium chloride in water

12. In an automobile battery, what substance makes up the plates at the cathode?

    a) PbO_2
    b) $PbSO_4$
    c) MnO_2
    d) Pb

13. What is the main disadvantage of molten carbonate fuel cells (MCFC)?

   a)   low efficiency
   b)   dependence on methane
   c)   toxic emissions
   d)   very expensive

14. Identify the oxidizing agent in the following chemical reaction from the given list.

$$Al + Cr^{3+} \rightarrow Al^{3+} + Cr$$

   a) Al^{3+}
   b)   Cr
   c)   Al
   d) Cr^{3+}

15. Which one of the following reactions involves the loss and gain of electrons between elements?

   a) $CO_2 + H_2O \rightarrow H_2CO_3$
   b) $NCl + NaOH \rightarrow NaCl + H_2O$
   c) $CaO + SO_3 \rightarrow CaSO_4$
   d) $2\,CO + O_2 \rightarrow 2\,CO_2$

16. What is the purpose of the porous plug in an electrochemical cell?

   a)   produce both cations and anions
   b)   allow electrons to flow
   c)   maintain a charge balance
   d)   maintain a constant temperature

17. In the final step of respiration, what substance is oxidized?

   a)   oxygen
   b)   water
   c)   glucose
   d)   carbon dioxide

ANSWERS TO SELF-TEST QUESTIONS

1.   CO

2.

$$Mg\!: \; + \; :\!\ddot{C}l\!-\!\ddot{C}l\!: \; \longrightarrow \; \left[:\!\ddot{C}l\!:\right]^{-} \; Mg^{2+} \; \left[:\!\ddot{C}l\!:\right]^{-}$$

3.   Al – oxidized, Mn – reduced

4.   oxidation

5. $3\,H_2 + N_2 \rightarrow 2\,NH_3$

6.   Anode (-), cathode (+)

7. $2\,H_2 + 4\,OH^- \rightarrow 4\,H_2O + 4\,e^-$

8. $2\,Fe + O_2 + 2\,H_2O \rightarrow Fe(OH)_2$

9.   c

10.   a

11.   d

12.   a

13.   b

14.   d

15.   d

16.   c

17.   c

CHAPTER 15

THE CHEMISTRY OF HOUSEHOLD PRODUCTS

ANSWERS TO QUESTIONS:

1. An orange contains polar molecules, such as citric acid, which mix well with the other polar molecules. Since water is polar it dissolves the acid molecules, removing them easily when you rinse. On the other hand, french fries are coated with grease. The molecules composing grease are nonpolar and do not mix with polar water, therefore the grease does not dissolve and is not washed away with water alone. However, soap molecules can be used to remove grease from the hands because they have both polar and nonpolar ends. As soapy water is applied to the greasy hands, the soap molecules are attracted to both water and grease, thus removing the grease from the hands.

3. Soaps and detergent

$$CH_3CH_2CH_2CH_2CH_2CH_2CH_2CH_2CH_2CH_2CH_2CH_2CH_2CH_2CH_2\overset{\overset{O}{\|}}{C}-O^-Na^+$$

    molecules,   hydrocarbon tail (nonpolar)   ionic head
    when added to water, aggregate near the   (polar)
    surface where they can best accommodate their dual nature. At the surface, the ionic head submerges into the water while the nonpolar tail protrudes up out of the water, thus the name surfactants.

5. A colloidal suspension is a mixture of one substance dispersed in a finely divided state throughout another. In a normal solution molecules mix on a molecular scale, but in a colloidal suspension the molecules clump together to form small particles within the other substance. Shining a light through a colloidal suspension will show a dispersion of the light beam by the particles. If this same beam of light was shone into clear water, it could not be seen.

7. Certain ions in water, such as magnesium and calcium, reduce the effectiveness of soap. These ions react with soap molecules to form a slimy, gray scum called curd that deposits on skin when bathing or on the sides of the bathtub producing that undesirable ring around the tub.

9. ABS detergents accumulate in the environment, causing suds to form in natural waters. LAS detergents are biodegradable and decompose over time forming CO_2, H_2O and SO_4^{2-}, all common substances in the environment.

11. Recent trends in cleaning laundry have favored a combination of anionic LAS surfactants with nonionic surfactants.

13. Sodium tripolyphosphate is no longer used in detergent additives because the presence of phosphates in waste water causes eutrophication that threatens marine life.

15. A combination of builders including sodium carbonate (Na_2CO_3) and zeolites has replaced tripolyphosphate in U.S. laundry formulations.

17. Solid NaOH is the main ingredient in drain cleaners. When added to water, the solid sodium hydroxide dissolves releasing large amounts of heat and producing a hot NaOH solution. This solution melts the grease contained in clogs and converts some of the grease molecules to soap molecules. These in turn can help dissolve more grease. The NaOH solution also dissolves protein strands in hair, a component of clogged pipes. Some drain cleaners also contain small bits of aluminum. These aluminum bits dissolve in the NaOH solution and produce hydrogen gas. The bubbling action helps to physically dislodge clogs.

19. The shape of hair is determined by three types of interactions between protein chains: hydrogen bonds, disulfide linkages, and salt bridges.

21. High pHs (very basic) will damage the salt bridges in hair because H^+ ions are removed from $-NH_3^+$ groups resulting in neutral $-NH_2$. Since $-NH_2$ has no positive charge it will not bond to $-COO^-$ and the salt bridge is broken.

23. Shampoos contain synthetic surfactants such as sodium lauryl sulfate, which remove sebum and do not form curd with hard water ions as some types of soap do. Shampoos also contain thickeners, suds boosters and fragrances.

25. The natural color of hair depends primarily on the presence of two pigments: melanin, a dark pigment, and phaeomelanin, a red-orange pigment.

27. Too little moisture causes excessive flaking of dead cells, while too much moisture leads to increased risk of infection by microorganisms.

29. Sunscreens are lotions that contain chemical compounds that absorb ultraviolet radiation. SPF or skin protection factor indicates the ability of a sunscreen to absorb ultraviolet light. The number, for example SPF-15, means that 15 hours in the sun with the sunscreen produces an exposure equivalent to one hour without it.

31. Facial foundations are oil and water emulsions with a soap-like emulsifying agent. Facial powders contain talc, calcium carbonate, and pigments and dyes for color. Mascara is composed of soap, oils, waxes and pigments.

33. Both animal fragrances, such as leather, and plant fragrances, such as jasmine, are used in perfumes. Chemists now synthesize many of the compounds responsible for plant and animal smells.

35. Bacteria, dwelling in warm and moist areas of the body, convert some of the organic compounds present in perspiration into compounds such as ammonia, hydrogen sulfide, and other compounds with unpleasant odors.

37. Antiperspirants function by reducing the amount of perspiration that sweat glands produce.

39. Low-density polyethylene (LDPE) is the soft, flexible plastic used in garbage bags and shopping bags. High-density polyethylene (HDPE) is the dense, tough plastic used in bottles that hold liquids.

41. Teflon contains C–F bonds that are particularly strong and resistant to chemical attack. Consequently, Teflon is a strong and inert plastic material that resists sticking or degradation by other compounds.

43. a) Nylon is used in nylon stockings, pantyhose, cords and bristles.

   b) Polyethylene terephthalate (PET) is used in clothing, in packaging, as a waterproof coating on sails, as a base for magnetic video and audiotapes, and as a base for photographic film.

   c) Polycarbonate is often used in eyewear protection, as a scratch-resistant coating for
       eye glasses and as bulletproof windows.

   b) Polyisoprene is the polymer in rubber used in bike tires, auto tires, engine mounts, building foundations, and bridge bearings.

   e) Styrene-butadiene rubber (SBR) is used extensively to replace natural rubber in automobile tires.

SOLUTIONS TO PROBLEMS:

45. Sodium palmitate

$$CH_3-(CH_2)\overline{_{14}}-\overset{\overset{\displaystyle O}{\|}}{C}-O^-Na^+$$

47.

$$
\begin{array}{c}
\mathrm{O} \\
\parallel \\
-\mathrm{C}-\mathrm{O}^- \cdots\ ^+\mathrm{H_3N}-
\end{array}
\qquad \text{Salt bridge}
$$

Hydroxide ions (OH⁻) are in large concentrations in high pH solutions. Yes, they would damage the salt bridge

49. Since toilet bowl cleaners are acids another household product that would work is vinegar, which is acetic acid, CH_3COOH. The reaction of acetic acid with calcium carbonate is as follows:

$$2\ CH_3COOH\ +\ CaCO_3\ \longrightarrow\ Ca(CH_3COO)_2\ +\ H_2CO_3$$

51. When the relative humidity is high the actual water vapor in the air is close to the maximum water vapor the air can hold. Sweat does not evaporate easily because the air is already saturated with water vapor.

53. A perfume that lacks a bottom note needs large compounds with high molecular weights added so that they will evaporate slowly making the perfume last longer on your skin.

55.　$-CH_2CH_2CH_2CH_2CH_2-$　　　　$-CH_2CH\cdot CH_2CH\cdot CH_2-$

　　　　polyethylene　　　　　　　　　　　　Cl　　　Cl

　　　　　　　　　　　　　　　　　　　　polyvinylchloride

57.

$$
\left[\begin{array}{c}
\mathrm{CH_3} \\
| \\
-\mathrm{CH_2}-\mathrm{C}- \\
| \\
\mathrm{CH_3}
\end{array}\right]_n
$$

59.

$$
\left[\begin{array}{c}
\mathrm{O}\quad\quad\mathrm{O} \\
\parallel\quad\quad\parallel \\
-\mathrm{NH}-\mathrm{C}\cdot(\mathrm{CH_2})_2-\mathrm{C}-\mathrm{NH}-(\mathrm{CH_2})_6-
\end{array}\right]_n
$$

131

SOLUTIONS TO FEATURE PROBLEMS:

63. A and C

REVIEW TESTS FOR CHAPTER 15

Match the following statements or phrases to the end-of-chapter terms.

1. The term applied to soap molecules that have aggregated in these structures.

2. A mixture of one substance dispersed in a finely divided state throughout another.

3. Colloidal solutions that consist of two substances that would not normally mix.

4. Type of detergents in which the polar part of the molecule is a negative ion.

5. Any substance that increases the effectiveness of a surfactant.

6. The process in which phosphates enrich algae in fresh or salt water.

7. A compound or substance that acts at the surface of a liquid.

8. Substances that help decompose tough stains on laundry.

9. The oily substance secreted by glands in the scalp.

10. The measure of the actual water content of air relative to its maximum water content.

11. Chain-like molecules composed of monomers.

12. A plastic that softens when heated and hardens when cooled.

13. This kind of polymer is composed of alternating units.

14. Polymers that expel atoms, usually water, during their formation.

15. A polymer that stretches easily and returns to its original shape.

16. The process of making rubber harder and more elastic by cross-linking polyisoprene chains with sulfur atoms.

17. Small porous particles used to soften water.

18. Substances added to detergents to increase bulk and control consistency and density.

ANSWERS TO MATCHING

1. micelles
2. colloidal suspension
3. emulsions
4. anionic detergents
5. builder
6. eutrophication
7. surfactant
8. bleaching agent
9. sebum
10. relative humidity
11. polymers
12. thermoplastic
13. copolymers
14. condensation polymers
15. elastomer
16. vulcanization
17. zeolites
18. fillers

SELF-TEST QUESTIONS

Completion: Write the word, phrase, or number in the blank which will complete the statement or answer the question.

1. What part of the soap molecule interacts with grease or oil?

2. Brighteners, when added to laundry formulations, absorb _____ light and emit _____ light.

3. Some toilet bowl cleaners contain HCl, which dissolves $CaCO_3$ that can develop in the toilet bowl.   Write the balanced chemical equation for this reaction.

133

4. Draw the structure of the monomer from which the following addition polymer is formed:

$$\left(CH_2 - \underset{\underset{Cl}{|}}{CH} \right)_n$$

5. Draw the structure of the condensation copolymer made from the following monomers:

   $HOOCCH_2COOH$ and $H_2N\ (CH_2)_2NH_2$

6. The reaction of carbonic acid with compounds such as bisphenol A produce condensation copolymers called _____.

Multiple Choice: (Select the correct answer from the choices listed.)

7. All of the following are polymers <u>except</u>

   a) teflon
   b) rubber
   c) propylene
   d) orlon

8. Polyoxyethylene is an example of a(n) _____ detergent.

   a) anionic
   b) cationic
   c) non-ionic

9. High pH's will damage salt bridges between keratin strands because H^+ ions are removed from _____ groups.

   a) $-NH_2$
   b) $-COOH$
   c) $-NH_4^+$
   d) $-NH_3^+$

10. Which one of the following substances is <u>not</u> a hydrocarbon?

   a) orlon
   b) rubber
   c) polyisobutylene
   d) polystyrene

11. Which one of the chemical formulas listed is classified as a builder in laundry-cleaning formulations?

   a) H_2O_2
   b) $Al_2(OH)_4Cl_2$
   c) $Na_5P_3O_{10}$
   d) $Na_3BO_3 \bullet 3H_2O$

12. What is the function of quaternary ammonium compounds in shampoos?

   a) remove sebum from hair and scalp
   b) reduce the tendency of hair to tangle
   c) produce a permanent curl in hair
   d) add color to hair

13. The most important ingredient in underarm deodorants are _____.

   a) perfumes
   b) detergents
   c) emulsifying agents
   d) antibacterial agents

14. What is a typical use of low-density polyethylene?

   a) garbage bags
   b) bottles
   c) table tops
   d) pipes

15. What is a disadvantage of styrene-butadiene rubber compared to natural rubber?

   a) cannot vary amounts of constituents
   b) degrades faster
   c) more expensive
   d) reduced strength

16. Which one of the following compounds could be the monomer of an addition polymer?

   a) trichloromethane
   b) 1,2-dichloroethene
   c) tetrachloromethane
   d) hexachloroethane

17. What is the purpose of adding talc to face powders?

   a) maintain moisture content
   b) absorb UV light
   c) absorb oil and water
   d) enhance brightness

ANSWERS TO SELF-TEST QUESTIONS

1. hydrocarbon or non-polar part
2. UV, visible
3. $2HCl + CaCO_3 \rightarrow CaCl_2 + H_2O + CO_2$
4. $CH_2=CHCl$
5.

$$\left((CH_2)_2NH - \overset{\overset{O}{\|}}{C}CH_2\overset{\overset{O}{\|}}{C} - NH \right)_n$$

6. polycarbonates
7. c
8. c
9. d
10. a
11. c
12. b
13. d
14. a
15. d
16. b
17. c

CHAPTER 16

BIOCHEMISTRY AND BIOTECHNOLOGY

ANSWERS TO QUESTIONS:

1. **Lipids** function as long-term energy storage and are the primary means for structural components of biomembranes

   Carbohydrates function as short-term energy storage.

   Proteins function as structural components for muscle, hair and skin, and control much of the body's chemistry.

   Nucleic acids function as information storehouses used to create protein.

3. Lipids are those cellular components that are insoluble in water, but extractable with non-polar solvents. A typical triglyceride has the following general structure:

$$
\begin{array}{c}
\quad\quad\quad\quad \overset{\displaystyle O}{\overset{\|}{}} \\
CH_2-O-C-R \\[4pt]
\quad\quad\quad\quad \overset{\displaystyle O}{\overset{\|}{}} \\
CH-O-C-R \\[4pt]
\quad\quad\quad\quad \overset{\displaystyle O}{\overset{\|}{}} \\
CH_2-O-C-R
\end{array}
$$

   R groups = saturated or unsaturated hydrocarbons.

5. Carbohydrates are a class of compounds derived from **carbo** meaning carbon, and **hydrate** meaning water. Their formulas are approximated by multiples of the formula, CH_2O, or simply $(C \cdot H_2O)_n$.

7. Carbohydrates are polyhydroxy aldehydes or ketones or their derivatives.

9. The large number of hydroxy (–OH) groups on carbohydrates allow for hydrogen bonding with each other and with water, making carbohydrates soluble in water and in bodily fluids.

11. A monosaccharide (glucose) contains only one carbohydrate unit. A disaccharide (sucrose) contains two monosaccharide units. Polysaccharides are complex carbohydrates (starch or cellulose) that contain many monosaccharide units.

137

13. Proteins are molecules composed of long chain repeating units called amino acids. Proteins function as structural materials in much of our muscle, hair, and skin and act as enzymes that regulate countless chemical reactions. They also serve as hormones to regulate metabolic processes, transport oxygen from the lungs to cells, and act as antibodies to fight invading organisms and viruses.

15. Twenty different amino acids exist in human proteins. They differ only in their **R** groups on the central carbon. A general structure for an amino acid is shown below:

$$H_2N-\overset{\overset{\displaystyle H}{|}}{\underset{\underset{\displaystyle R}{|}}{C}}-\overset{\overset{\displaystyle O}{||}}{C}-OH$$

These different **R** groups could be polar, non-polar, acidic, basic, or neutral in character. Thus, the water solubility or water insolubility and pH in aqueous solutions of the various amino acids are controlled by these **R** groups.

17. The secondary structure of a protein refers to the way the chain of amino acids orients itself along its axis. This secondary structure is held together by hydrogen bonding between the oxygen (see fig. A) and hydrogen (see fig. B) groups in the polypeptide chains.

Fig A
$$\begin{array}{c} \backslash \\ C=O \\ / \end{array}$$

Fig. B
$$\begin{array}{c} / \\ H-N \\ \backslash \end{array}$$

19. Hemoglobin is a protein composed of four polypeptide chains arranged in the proper configuration to hold four flat molecules called heme groups. These heme groups bind readily with oxygen in the lungs and carry it to cells where it is needed for glucose oxidation.

21. Lysozyme is a protein, which functions as an enzyme. An enzyme is a substance that catalyzes or promotes a specific chemical reaction. It was discovered by Alexander Fleming who placed nasal mucus on a dish containing bacteria. The mucus was able to dissolve away the bacteria.

23.

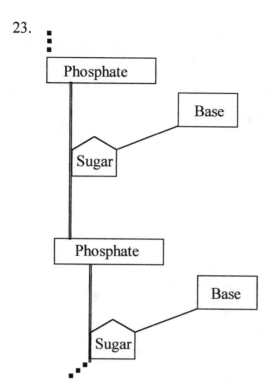

25. DNA contains four bases: adenine, guanine, cytosine, and thymine. The complementary bases are adenine-thymine and guanine-cytosine.

27. Chromosomes are biological structures that contain DNA material. Human beings have 46 chromosomes, found in the nucleus of a cell.

29. The order of codons along the DNA molecule determines the order of amino acids in a protein. A section of DNA that codes for one protein is called a gene. However, not all of the DNA is expressed in each cell. The expression of DNA requires the formation of m-RNA followed by protein synthesis at the ribosome. Cells only express the proteins specific to their function.

31.

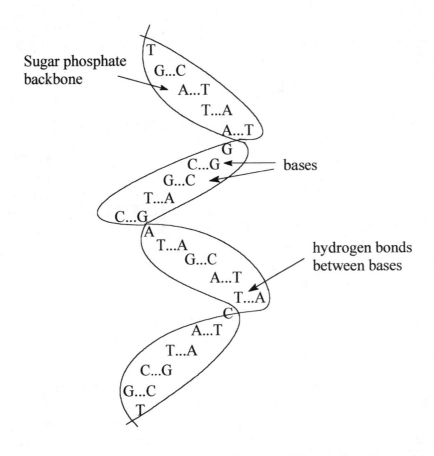

Sugar phosphate
backbone

bases

hydrogen bonds
between bases

33. Recombinant DNA is the result of a transfer of DNA from one organism to another. The combination of the organisms original DNA and the DNA transferred is called recombinant DNA.

35. Scientists can now use genetic engineering to isolate a gene that codes for the production of a specific protein. The gene is transferred into bacteria, which is then grown in a culture. As the bacteria multiplies so does the foreign DNA resulting in millions of copies of the desired protein. The proteins now being genetically engineered include human insulin and human growth hormone.

37. DNA can be isolated from blood, semen, or most any bodily fluid or tissue. The DNA can then be screened for genes that may increase a person's chances of developing certain diseases such as cancer, heart disease, and Alzheimer's. By knowing early on that they are predisposed to certain diseases, a person can take preventative steps to avoid it.

39. In DNA fingerprinting, a small sample of a person's blood, hair, skin, semen or other bodily tissue or fluid is collected at a crime scene. The DNA within the cells from the

140

sample is isolated and cut into fragments using restriction enzymes. These DNA fragments are then separated by gel electrophoresis and a radioactive probe, which attaches to a specific base sequence, is then applied to the mixture. The radioactive sample is allowed to expose a photographic film that produces a pattern of light and dark. Since every person's DNA is different, the light and dark bands are unique and are called a DNA fingerprint. This procedure is useful for a trial because a positive match between the DNA fingerprints of the suspect and a sample from the crime scene links the suspect to the crime.

41. Cloning has been accomplished with a fertilized human egg. The human egg was allowed to divide into two and then the two cells were separated. It was revealed that each cell continued to divide in a normal fashion to produce genetically identical embryos. However, the embryos were intentionally defective and did not develop any further.

SOLUTIONS TO PROBLEMS:

43. compound (d)

45. compound (b)

47. compound (a)

49.

51. a) monosaccharide
   b) polysaccharide
   c) disaccharide
   d) disaccharide

53. compound (b)

55. a) carbohydrate
   b) lipid
   c) lipid

d) amino acid

57.

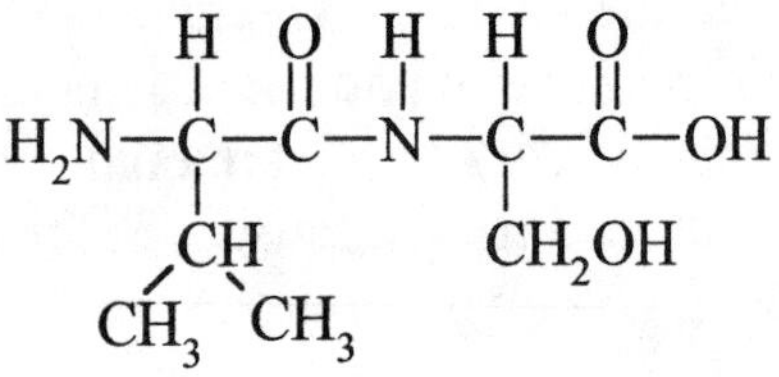

$$H_2N-\underset{\underset{\underset{CH_3}{CH_3}}{\overset{CH}{|}}}{\overset{\overset{H}{|}}{C}}-\underset{}{\overset{\overset{O}{\|}}{C}}-\underset{}{\overset{\overset{H}{|}}{N}}-\underset{\underset{CH_2OH}{|}}{\overset{\overset{H}{|}}{C}}-\underset{}{\overset{\overset{H\;\;O}{\;\;\;\|}}{C}}-OH$$

59.

$$H_2N-\underset{\underset{CH(CH_3)_2}{\overset{CH_2}{|}}}{\overset{\overset{H}{|}}{C}}-\underset{}{\overset{\overset{O}{\|}}{C}}-\underset{}{\overset{\overset{H}{|}}{N}}-\underset{\underset{CH(CH_3)_2}{\overset{CH_2}{|}}}{\overset{\overset{H}{|}}{C}}-\underset{}{\overset{\overset{O}{\|}}{C}}-\underset{}{\overset{\overset{H}{|}}{N}}-\underset{\underset{CH(CH_3)_2}{\overset{CH_2}{|}}}{\overset{\overset{H}{|}}{C}}-\underset{}{\overset{\overset{O}{\|}}{C}}-OH$$

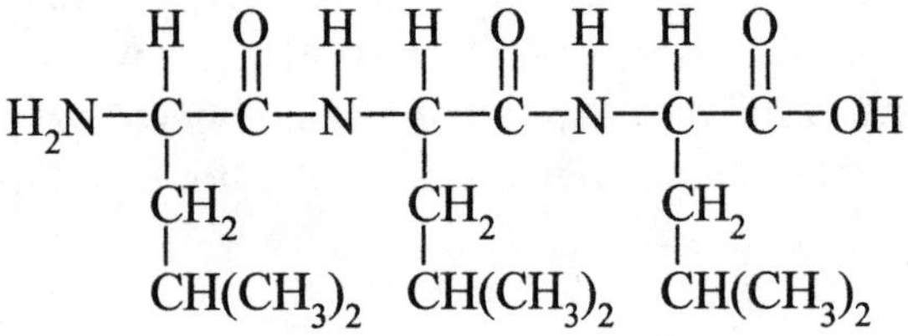

You would not expect this tripeptide to mix readily with water because of its non-polar hydrocarbon side chains. These groups give the peptide a hydrophobic nature.

61. a) thymine
   b) guanine
   c) cytosine
   d) adenine

SOLUTIONS TO FEATURE PROBLEMS:

69. a) amino acid
   b) monosaccharide
   c) disaccharide
   d) amino acid

REVIEW TESTS FOR CHAPTER 16

Match the following statements or phrases to the end-of-chapter terms.

1.   The study of the application of biochemical knowledge to living organisms.

2.   The primary means for short-term energy storage in living organisms.

3.   Substances that provide the information to create proteins.

4.   An organic acid with a long hydrocarbon tail.

5.   The combination of glycerol with three fatty acids.

6.   Cellular components that are insoluble in water, but extractable with non-polar solvents.

7.   The amino acid sequence of a protein.

8.   A type of secondary structure in which a protein forms zigzag chains that stack neatly on top of each other to form sheets held together by hydrogen bonds.

9.   The type of structure in a protein that consists of two or more sub-chains that are themselves held together by interactions between the chains.

10.   A substance that regulates specific processes within the body.

11.   The two categories of nucleic acids.

12.   The repeating units that compose nucleus acids.

13.   A group of three bases along a nucleic acid chain.

14.   The structure where protein synthesis occurs.

15.   The type of enzymes that cut DNA at specific places along the DNA strand.

16.   A term used to describe monosaccharides that join in long chains.

17.   A sugar consisting of two monosaccharide molecules joined together.

18. A technique in which a mixture of DNA fragments is placed on a thin gel and then an electric current is applied that causes the mixture to separate into distinct bands.

ANSWERS TO MATCHING

1.   biotechnology
2.   carbohydrates
3.   nucleic acids
4.   fatty acid
5.   triglyceride
6.   lipids
7.   primary structure
8.   pleated sheet
9.   quaternary structure
10. proteins
11. RNA and DNA
12. nucleotides
13. codon
14. ribosome
15. restrictive enzymes
16. polysaccharides
17. disaccharide
18. gel electrophoresis

SELF-TEST QUESTIONS

<u>**Completion:**</u> Write the word, phrase, or number in the blank which will complete the statement or answer the question.

1.   The substances that control most of our body's chemistry are called
_____.

2.   Fats and oils would be part of what class of biochemical compounds?

3.   Draw the structure for the straight chain form of glucose.

4.   Draw the general structure for an amino acid.

5.   Draw the complementary strand for the following DNA fragment.

$$\overline{}$$
A   T   G   C

6.   Classify the following fatty acid R group as saturated or unsaturated. Would it form a fat that is a solid or liquid at room temperature?

$$\overset{\displaystyle O}{\overset{\displaystyle \|}{-C(CH_2)_{16}CH_3}}$$

7.   What do the letters in DNA stand for?

8.   How many chromosomes exist in humans?

9.   Each of the four bases A, T, C and G, has a complementary partner with which it _____ bonds.

10. In starch, the oxygen atoms joining the glucose units points down relative to the planes of the rings and are called _____ linkages.

Multiple choice: Select the correct answer from the following choices.

11. Which one of the following substances would be classified as a carbohydrate?

     a)   triolein
     b)   cellulose
     c)   adenine
     d)   α-keratin

12. The alpha-helix is part of the _____ structure of proteins.

     a)   primary
     b)   secondary
     c)   tertiary
     d)   quaternary

13. Hemoglobin would be categorized as a

   a) nucleic acid
   b) carbohydrate
   c) lipid
   d) protein

14. A nucleotide consists of all of the following components except a(n)

   a) amino acid
   b) phosphate group
   c) sugar group
   d) base

15. In sucrose, a _____ unit and a _____ unit are linked together to form a two-ringed structure.

   a) glucose, glucose
   b) fructose, fructose
   c) glucose, fructose
   d) raffinose, fructose

16. What is the biochemical classification for the following structure?

$$HOCH_2 \quad O \quad OH$$
$$HO$$
$$CH_2OH$$
$$OH$$

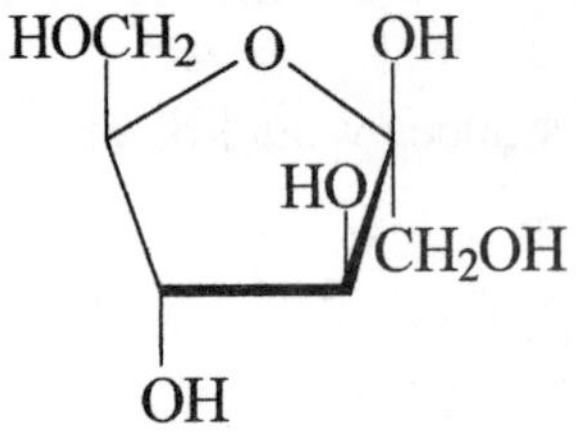

   a) protein
   b) lipid
   c) nucleic acid
   d) carbohydrate

17. For the following structure, which numbered arrow correctly identifies the polar part?

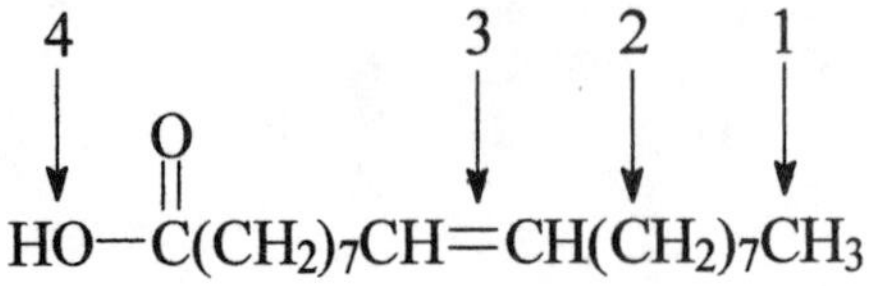

4   3   2   1

$$HO-\overset{\overset{\textstyle O}{\|}}{C}(CH_2)_7CH=CH(CH_2)_7CH_3$$

a) 1
b) 2
c) 3
d) 4

18. What type of peptide, if any, does the following structure indicate?

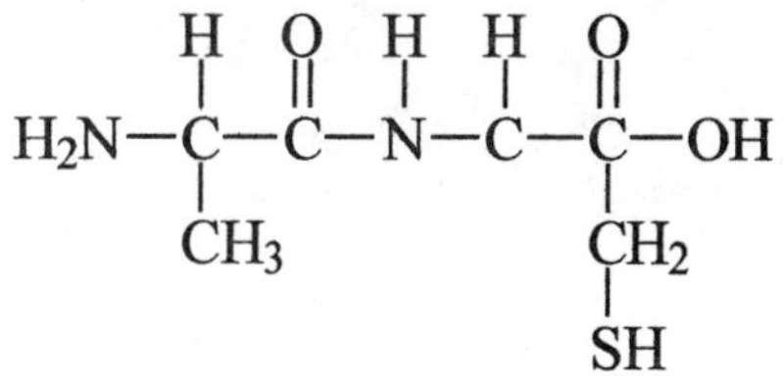

$$H_2N-\overset{\overset{\displaystyle H}{|}}{\underset{\underset{\displaystyle CH_3}{|}}{C}}-\overset{\overset{\displaystyle O}{||}}{C}-\overset{\overset{\displaystyle H}{|}}{N}-\overset{\overset{\displaystyle H}{|}}{C}-\overset{\overset{\displaystyle O}{||}}{\underset{\underset{\underset{\underset{\displaystyle SH}{|}}{CH_2}}{|}}{C}}-OH$$

a) dipeptide
b) tripeptide
c) tetrapeptide
d) the structure does not represent a peptide

19. What two kinds of interactions gives hair its shape?

a) ionic bonds and sulfur bridges
b) sulfur bridges and nitrogen bridges
c) hydrogen bonds and sulfur brides
d) hydrogen bonds and nitrogen bridges

20. What is the term used to describe the new DNA that results from the transfer of DNA from one organism to another?

a) restriction DNA
b) recombinant DNA
c) transfer DNA
d) ribosomal DNA

ANSWERS TO SELF-TEST QUESTIONS

1. proteins
2. lipids

147

3.

$$\begin{array}{c}
H-C{=}O \\
| \\
H-C-OH \\
| \\
HO-C-H \\
| \\
H-C-OH \\
| \\
H-C-OH \\
| \\
CH_2OH
\end{array}$$

4.

$$\begin{array}{c}
H \\
| \\
R-C-COOH \\
| \\
NH_2
\end{array}$$

5.

T A C G

6. saturated; solid
7. deoxyribonucleic acid
8. 46
9. hydrogen
10. alpha
11. b
12. b
13. d
14. a
15. c
16. d
17. d
18. a
19. c
20. b

148

CHAPTER 17

DRUGS AND MEDICINE: HEALING, HELPING, AND HURTING

ANSWERS TO QUESTIONS:

1. Phenylethylamine is the molecule located in the brain that is associated with feelings of being in love.

3. Aspirin performs as an analgesic, anti-pyretic, and anti-inflammatory by preventing the formation of prostaglandins, fatty acid derivatives involved in a number of physiological processes.

5. Aspirin does have some relatively minor side effects; its acidity can irritate the stomach, it reduces the formation of blood platelets which initiate blood clotting, it is toxic in large doses, and it could cause Reye's Syndrome in children with chicken pox or flu-like symptoms.

7. The only differences between generically labeled drugs and their brand name counter part are the price and the names.

9. Antibiotics are drugs that fight bacterial diseases by targeting the unique physiology of bacteria and selectively killing them.

11. Penicillin and cephalosporin kill bacteria by destroying their cell walls. Once the cell wall is weakened the cell wall ruptures and the bacteria die. However, human cell walls are different, and are not affected by these compounds.

13. Antibiotics can lose their effectiveness because some bacteria have evolved to be resistant to drugs. Bacterial strains have emerged that are completely drug-resistant.

15. A retrovirus consists of RNA and protein whereas most viruses contain DNA and protein. Viral RNA enters the host cell, then is reversed transcribed into DNA. This DNA then goes on to direct the synthesis of viral protein.

17. AZT is a nucleoside analog, an antiviral drug. AZT is incorporated into DNA by viral enzymes that are fooled by the phosphorylated AZT, which resembles the nucleotide thymine. When the viral enzymes construct viral DNA, the analogue is incorporated and because it has only one point of attachment, the synthesis of the DNA is terminated. Cells infected with the AIDS virus prefer the fake or phony nucleotide to the real one, which slows the progression of the disease by stopping the production of the viral DNA. Healthy cells, on the other hand, do not show this preference. Unfortunately, mutant forms of the virus developed that showed no preference for the phony nucleotide. The result for patients is that taking AZT only slows the progression of AIDS for a short time period.

19. A powerful combination of drugs – a protease inhibitor, AZT, and another nucleoside analog – can kill the HIV at two different points in its replication cycle. In order for the virus to survive, it would have to develop resistance to both drugs simultaneously, an unlikely event.

21. Testosterone is a male sexual hormone, which acts before the birth of a male infant to form the male sex organs. Testosterone functions at puberty and beyond to promote the growth and maturation of the male reproduction system, to develop the sexual drive and to determine secondary sexual characteristics such as facial hair and voice deepening.

23. The birth control pill contains synthetic analogs of the two female sex hormones. The estrogen analog regulates the menstrual cycle while the progesterone analog establishes a state of false pregnancy. Taking these hormones on a daily basis allows for the maintenance of a false pregnancy state, therefore no eggs are released, preventing conception. Nausea, weight gain, fluid retention, breast tenderness, and acne are all possible side effects of the birth control pill.

25. Steroids are chemical substances characterized by their seventeen-carbon-atom, four-fused-ring, skeletal structure, as shown below.

Three examples of steroids are progestin, cortisone, and prednisone.

27. Anabolic steroids act to increase muscle mass. However, these steroids have serious side effects; in males, lower sperm production and testicular atrophy are observed. While in females, male secondary sexual characteristics, irregular menstruation, and ovulation failure are possible side effects. In both sexes, anabolic steroids increase the risk of liver damage, stroke, and heart attack.

29. The side effects of chemotherapy include anemia, the inability to fight infection, nausea, vomiting, infertility, hair loss, fatigue, and internal bleeding.

31. a) Alkylating agents are highly reactive compounds that add alkyl groups (e.g. $-CH_2CH_3$) to other organic or biological molecules. The alkylation of DNA within cells produces defects in the DNA that results in cell death.

   b) Antimetabolites are chemical substances that function by impersonating compounds normally found in the cell, but possess subtle differences that usually stop DNA synthesis.

c) Topoisomerase inhibitors function by altering the action of a topoisomerase, an enzyme that pulls apart the individual strands of DNA in preparation of DNA synthesis. The topoisomerase inhibitor results in DNA damage, and thus, cell death eventually occurs. These compounds can cause cells to die whenever they initiate replication.

d) Hormone treatment involves those cancers involving the breasts or sexual organs. Cancer cells within the tissues need a large dose of hormones to grow. Denying the tissue the needed hormones stops cancerous tumor growth.

33. The body metabolizes ethanol in the liver where it expels the ethanol equivalent of one drink per hour.

35. Alcoholics often drink in the morning, become depressed due to lack of drinking, drink alone or in secret, and may suffer blackouts or memory loss after episodes of heavy drinking.

37. Some common depressant inhalants used for anesthetic purposes are nitrous oxide (laughing gas), isoflurane, halothane, and enflurane.

39. The immediate effects of inhaling vapors of household chemicals can include alcohol-like intoxication, loss of consciousness, and even death. Long term problems of vapor inhalation are brain damage, kidney damage, and liver damage.

41. Narcotics function by binding to opioid receptors, specific sites on nerve cells in the brain and spinal cord. The narcotic molecule, when bound to an opioid receptor, inhibits certain neurotransmitters involved in the transmission of pain impulses.

43. Codeine is less potent than morphine only effective in the relief of moderate pain. Also, its tolerance develops more slowly and it is less likely to produce addiction, compared to morphine. Heroin is three times more potent than morphine and also extremely addictive. Its effect, compared to morphine, is much more intense, since it travels to the brain faster.

45. Naltrexone binds to opioid receptor sites, displacing opioids, without producing the narcotic effects. When naltrexone is taken daily, any ingestion of an opioid narcotic will be ineffective. Addicts can use this drug as part of their recovery program. The craving for the narcotic is partly overcome by the knowledge that the drug will not produce the desired high.

Methadone works in the same way as Naltrexone, binding to opioid receptor sites. Unlike other narcotics, it does not produce stupor. However, methadone does prevent the physical symptoms associated with withdrawal.

47. Stimulants are drugs that increase the activity of the central nervous system. Examples of stimulants are caffeine, amphetamines, nicotine, and cocaine.

49. Caffeine, once in the bloodstream, interacts with nerve cells to block adenosine receptors. The result is an increase of dopamine and norepinephrine neurotransmitters. The effects of increased levels of these neurotransmitters are a feeling of alertness, competence and wakefulness.

51. Nicotine interacts with acetylcholine receptors in the central nervous system. This interaction result in a faster heart rate, elevated blood pressure, and causes the release of adrenaline, promoting a fight-or-flight response.

53. Hallucinogenic drugs are chemical substances that distort and disturb cognition and perception. Mescaline and LSD are examples of hallucinogenic drugs.

55. Lysergic Acid Diethylamine (LSD) was first synthesized by Dr. Albert Hoffman, a Swiss chemist, in 1938. Dr. Hoffman took several intentional doses of LSD after unintentionally experiencing its hallucinogenic effects while working with the drug. He documented his results and his records indeed revealed the hallucinogenic powers of LSD.

57. LSD can produce rapid changes in mood and emotion as well as severe perceptual distortions and hallucinations. Users experience "seeing sounds" or "hearing colors" along with a significant slowing in the passage of time. Chronic use of LSD can lead to a lower capacity for abstract thinking as well as permanent brain damage.

59. Low doses of marijuana produce relaxation, relief from anxiety, increases sense of well-being, alteration in sense of time and mild euphoria. As the dosage increases the effects increase from mild hallucinations to depression, panic attacks, and paranoia. The side effects of taking marijuana are controversial; short term effects appear to be minimal with only a mild physical dependency. The long term effects though are more severe and are the result of THC build up in the body. The effects include interference with cognitive function, and learning, reasoning and academic impairments. Other long-term effects include lung and throat irritation, increased risk of lung cancer, and partial suppression of the immune system.

61. Since a deficiency in the neurotransmitter serotonin within the brain appears to cause depressions, both Prozac and Zoloft function by elevating serotonin levels back to normal. This results in reversing the adaptive changes caused by low levels of serotonin and relieving the depression. The side effects of Prozac include nervousness, dizziness, anxiety, sexual dysfunction, insomnia, nausea and loss of appetite. The side effects of Zoloft include dizziness, sexual dysfunction, insomnia, nausea, diarrhea and dry mouth.

SOLUTIONS TO PROBLEMS:

63.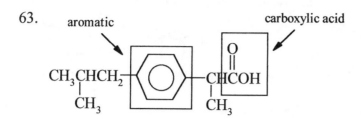

65. $75 \text{ mg} \times \dfrac{1 \text{ mL}}{20 \text{ mg}} = 3.75 \text{ mL}$

67. $80 \text{ kg} \times \dfrac{17.5 \text{ mg}}{1 \text{ kg}} = 1400 \text{ mg or } 1.4 \text{ g}$

SOLUTIONS TO FEATURE PROBLEMS:

73. a) $80 - 76 = 4 \text{ yrs}$

    b) $\dfrac{4 \text{ yrs}}{17 \text{ yrs}} = 0.24 \text{ yr increase per year}$

    c) $2030 - 1997 = 33 \text{ years}$

       $33 \text{ yr} \times \dfrac{0.24 \text{ yr}}{1 \text{ yr}} = 7.9 \text{ year increase from the life expectancy in 1997}$

       $80 + 7.9 = 88 \text{ yrs}$

REVIEW TESTS FOR CHAPTER 17

Match the following statements or phrases to the end-of-chapter terms.

1.   A substance that helps transmit messages among nerve cells.

2.   These compounds stimulate the release of prostaglandins.

3.   These chemical substances mediate inflammation.

4.   Microorganisms that can reproduce within the human body.

5.   These type of viruses consist of RNA and protein.

6.   The name of the antibacterial agent discovered by Alexander Fleming.

7.   These enzymes act like molecular scissors to cut viral proteins to the correct size.

8.   The hormone that promotes the growth and maturation of the male reproductive system.

9.   The class of compounds characterized by their seventeen-carbon-atom, four-ring, skeletal structure.

10. The class of drugs used to treat narcotic overdoses.

11. Those drugs that act on the central nervous system to produce an analgesic and sedative effect.

12. The primary component of opium and was isolated from the opium poppy sap.

13. Drugs such as Valium and Librium fall under this type of compound.

14. The alcohol found in alcoholic drinks.

15. These chemical substances work by mimicking compounds normally found in the cell, but have subtle differences that usually stop DNA synthesis.

16. The active drug found in tobacco products.

17. The most abused amphetamine today, also known as speed.

18. Those class of drugs that distort and disturb cognition and perception.

19. A synthetic opioid drug that binds to opioid receptor sites.

20. A class of antiviral drugs that work by fooling the enzymes that construct viral DNA)

ANSWERS TO MATCHING

1.   neurotransmitter
2.   pyrogens
3.   histamines
4.   bacteria
5.   retroviruses
6.   penicillin

7. proteases
8. testosterone
9. steroids
10. opioid antagonist
11. narcotics
12. morphine
13. benzodiazapenes
14. ethanol
15. anti-metabolites
16. nicotine
17. methamphetamine
18. hallucinogenic drugs
19. methadone
20. nucleoside drugs

SELF-TEST QUESTIONS

Completion: Fill in the blank or provide an answer for each of the following.

1. Aspirin reduces pain and lowers fever by preventing the formation of
   _____.

2. What are the two female sex hormones secreted primarily by the ovaries?

3. Consider the structure of testosterone as show below.

   Identify the functional groups on this molecule.

4. When the compound cyclophosphamide enters a cell, it adds _____ groups
   to DNA, producing defects in DNA that result in cell death.

5. What are the class of compounds implicated in explaining the common effect of a
   "runner's high"?

6. A nurse is to administer 75 mg Demerol from a container marked 100 mg/2 mL.
   How many mL are required?

Multiple choice: (Select the correct answer from the choices listed.)

7. Feelings of love are associated with elevated levels of a molecule called
   _____ in the brain.

   a) phenylamine
   b) ethylamine
   c) acetylsalicylic acid
   d) phenylethylamine

8. Which antibiotic, if any, prevents the normal development of bacterial cells walls
   but does not inhibit protein synthesis?

   a) acromycin
   b) penicillin
   c) terramycin
   d) none of the antibiotics listed)

9. Ru-486 is a drug which works by

   a) blocking the action of progesterone.
   b) decreasing prostaglandin levels in the blood)
   c) decreasing sugar levels in the blood)
   d) lowering sperm production.

10. The deficit of what neurotransmitter is currently believed to cause clinical
    depression?

    a) adenosine
    b) norepinephrine
    c) serotonin
    d) dopamine

11. Which chemical group listed is not used in chemotherapy to fight cancer?

    a) topoisomerase inhibitors
    b) anti-metabolites
    c) opioid antagonists
    d) hormones

12. What is the new class of anti-viral drugs developed against HIV?

   a) protease inhibitors
   b) topoisomerase inhibitors
   c) cephalosporins
   d) benzodiazapenes

13. What class of drugs are the most physically addictive?

   a) depressants
   b) stimulants
   c) narcotics
   d) hallucinogenics

14. When AZT enters infected cells, they are phosphorylated and thus become _____ analogs.

   a) nucleotide
   b) nucleoside
   c) pyrogen
   d) protease

15. Anabolic steroids are usually synthetic analogs of

   a) estrogen
   b) cortisone
   c) testosterone
   d) prednisone

16. What type of drug is ethanol?

   a) stimulant
   b) depressant
   c) narcotic
   d) hallucinogenic

17. What is the difference between bacteria and viruses?

   a) Viruses respond to antibiotics but bacteria do not.
   b) Bacteria consist of only DNA and protein while viruses consist of many cells.
   c) Bacteria cause colds and the flu but viruses do not.
   d) Bacteria reproduce on their own while viruses require a host cell to reproduce.

ANSWERS TO SELF-TEST QUESTIONS

1. prostaglandins
2. estrogen and progesterone
3. alcohol, alkene, ketone
4. alkyl
5. endorphins

6.

$$75 \; \cancel{mg} \; \times \; \frac{2 \; mL}{100 \; \cancel{mg}} \; = \; 1.5 \; mL$$

7. d
8. b
9. a
10. c
11. c
12. a
13. c
14. c
15. c
16. b
17. d

CHAPTER 18

THE CHEMISTRY OF FOOD

ANSWERS TO QUESTIONS:

1. The atoms contained in human bodies come from food.

3. Carbohydrates are the starches, sugars, and fibers contained in food. Chemically, carbohydrates are polyhydroxy aldehydes or ketones or their derivatives.

5. *Monosaccharides* are the simplest carbohydrates; they are the unit building block for other more complex carbohydrates. *Disaccharides* are simple carbohydrates that contain two monosaccharide units, which can be hydrolyzed into their respective monosaccharide units. Polysaccharides are complex carbohydrates composed of many monosaccharide units (glucose, for example) linked together.

7. Monosaccharides are the basic carbohydrate unit and can be directly transported in the blood to cells where they are oxidized in a number of reaction steps. Disaccharides and polysaccharides (except fiber) are hydrolyzed with water at the bonds between monosaccharide units into their composite units, which are then passed into the bloodstream.

9. Both starch and fiber are composed of long chains of glucose units. However, in starch, the glucose units are linked together by alpha linkages while in fiber, the links are beta linkages. This difference in the structures results in starch being digestible, whereas, fiber is not. The intestinal enzymes recognize the alpha linkage and attack them, but not the beta linkages.

11. In a healthy diet, the FDA recommends that carbohydrates compose 60% of caloric intake with 10% or less of total caloric intake from simple sugar.

13. Proteins are broken down into their amino acid components in the digestive tract. The amino acid components are then absorbed through the intestinal wall and into the bloodstream.

15. Complete proteins are proteins that contain all the essential amino acids in the right proportion. By making the right combinations, vegetarians can obtain complete proteins in their diets. Some of these right combinations include beans and rice, peanut butter and bread, and rice with tofu.

17. Saturated fats are those containing only single bonds and tend to be solids at room temperature. Unsaturated fats contain double bonds and exist as liquids at room temperature.

19. Excessive amounts of cholesterol in the blood can result in the excessive deposition of cholesterol on arterial walls, leading to a blockage of the arteries, a condition called arteriosclerosis. Such blockages are dangerous because they inhibit blood flow to important organs such as the heart or brain. The risk of stroke and heart attack increases with increasing blood cholesterol levels.

21. The FDA recommends that fats compose less than 30% of total caloric intake.

23. Fats and oils contain over twice the number of calories per gram than carbohydrates or proteins. Fiber contains no caloric content.

25. Fat is the most efficient way to store energy because it has the highest caloric content.

27. The fat-soluble vitamins include vitamins A, D, E and K. The water-soluble vitamins are classified as the eight B vitamins and vitamin C. Fat-soluble vitamins are soluble in fatty tissue and not easily excreted and can, therefore, be over consumed.

29. In the body, vitamin C functions in the synthesis of connective tissue called collagen, in the protection against infections, and in the absorption of iron through the intestinal wall. Vitamin C deficiency results in a condition called scurvy, in which bodily tissues and blood vessels become weakened. This condition can result in massive bleeding and if untreated, can lead to death. Vitamin C is present in citrus fruits, green leafy vegetables, cantaloupe, strawberries, peppers, tomatoes, potatoes, papayas and mangoes.

31. The B complex vitamins play important roles in metabolism (the extraction of energy from food), protein synthesis, and cell multiplication. Deficiencies in the B complex vitamins result in the inability for cells to extract energy from foods. The symptoms of the deficiency include exhaustion, irritability, depression, forgetfulness, partial paralysis, abnormal heart action, and severe skin problems. Green leafy vegetables and in the hulls of grain are rich in the B complex vitamins.

33. Calcium serves as the main structural material for teeth and bones. It also plays an important role in the transmission of nerve signals and in blood clotting. Low calcium levels in the blood will result in osteoporosis, a condition in which bones become weakened by calcium loss. Milk, cheese, sardines oysters and broccoli are foods rich in calcium.

Phosphorus is involved in the structure of teeth and bones. It also assists in energy metabolism and is a component of DNA. The effects of deficiencies of phosphorus are unknown since phosphorus is contained in most diets.

Iodine is used in the synthesis of a hormone that regulates basal metabolic rate. Its deficiency results in goiter, a condition in which the thyroid gland swells as it tries to absorb as many iodine particles as possible. Iodine deficiency, if it occurs during pregnancy, can result in severe retardation of the developing baby. Iodine is readily available in seafood, milk and iodized salt.

Iron is a critical component of hemoglobin, the protein that carries oxygen in the blood. Hence, loss of iron occurs through bleeding and its deficiency results in the body's inability to make enough hemoglobin. This problem result in a condition called anemia in which a person feels tired, listless, and susceptible to cold temperatures. Meat, fish poultry, clams, legumes and green leafy vegetables are all foods that are rich in iron.

Zinc is essential to the function of more than 100 enzymes. It is also important in growth and development, immune function, learning, wound healing and sperm production. Vegetarians are most at risk for zinc deficiency, since plant sources contain only small amounts of zinc. Over consumption of zinc is toxic. Animal products are excellent sources of zinc; these include meat, shellfish and poultry.

35. Potassium and magnesium are both involved in maintaining electrolyte balance in and around cells. Magnesium also plays an important role in bone formation and in the operations of many enzymes including those involved in the release of energy. Potassium occurs in significant amounts in fresh fruits and vegetables including bananas, cantaloupe and lima beans. Magnesium is found in oysters, sunflower seeds, spinach, and in "hard" drinking water.

37. Additives are often added to foods to preserve it, enhance its flavor or color and maintain its appearance or physical characteristics. The Food and Drug Administration (FDA) monitors and regulates all food additives.

39.

Antimicrobial Agent	Function
Salt	Drying meat and fish
Sodium nitrite	Inhibits the growth of bacteria
Sodium benzoate	Prevents microbial growth in packaged foods

41. Sulfites and EDTA are both antioxidants. Sulfites function as preservatives and anti-browning agents. EDTA acts to immobilize metal ions that catalyze oxidation reactions.

43. Artificial flavors include sugar, corn syrup, aspartame, oil of wintergreen, peppermint, ginger, vanilla and almond extract. A typical flavor enhancer used in foods is monosodium glutamate (MSG). Artificial flavors and flavor enhancers are added to food to improve the taste.

45. The primary nutrients for plants are potassium, nitrogen and phosphorus.

47. The Haber Process in an industrial reaction used to fix nitrogen (i.e. ammonia synthesis).

49. The secondary nutrients for plants are calcium, magnesium, and sulfur. The micronutrients required for plants include boron, chlorine, copper, iron, manganese, molybdenum, sodium, vanadium and zinc; however, these nutrients are only needed in small quantities and rarely need to be replenished in soils.

51. Bioamplication is a process in which a chemical substance becomes concentrated as it moves up the food chain (aquatic plants $\rightarrow$ fish $\rightarrow$ birds).

53. Broad-spectrum insecticides, those of the organophosphate family, are able to attack a wide range of pests. Examples of broad-spectrum insecticides are malathion and parathion. Narrow-spectrum insecticides are of the carbamate type, such as carbaryl and aldicarb, and target only specific pests.

55.
Modern Herbicides	Uses
Atrazine	Destroys weeds in cornfields
Metolachlor	Used on soybeans and corn
Paraquat	Destroys marijuana crops

SOLUTIONS TO PROBLEMS:

57. Same calculation as above only you insert your weight to begin with. Answers will vary by person.

59.
$$\frac{6.0\,lbs}{1\,month} \times \frac{1\,month}{30\,days} \times \frac{-3500\,cal}{1\,lb} = -700\,cal\,/\,day$$

Since the person expends 2800 calories, the intake that would provide a –700 calorie deficit is:

$$2800\,Cal - 700\,Cal = 2100\,Cal\,/\,day$$

162

61. a) <u>Little Debbie Choco-Cakes</u>

Fat : $13\,\text{g} \times \dfrac{9\,\text{Cal}}{1\,\text{g}} = 117\,\text{Cal}$

Protein : $2\,\text{g} \times \dfrac{4\,\text{Cal}}{1\,\text{g}} = 8\,\text{Cal}$

Carbohydrate : $37\,\text{g} \times \dfrac{4\,\text{Cal}}{1\,\text{g}} = 148\,\text{Cal}$

Total : $273\,\text{Cal}$

The percentage of calories from fat :

$\dfrac{117\,\text{Cal}}{273\,\text{Cal}} \times 100\% = 43\,\%$

b) <u>Snickers Bar</u>

Fat : $14\,\text{g} \times \dfrac{9\,\text{Cal}}{1\,\text{g}} = 126\,\text{Cal}$

Protein : $4\,\text{g} \times \dfrac{4\,\text{Cal}}{1\,\text{g}} = 16\,\text{Cal}$

Carbohydrate : $35\,\text{g} \times \dfrac{4\,\text{Cal}}{1\,\text{g}} = 140\,\text{Cal}$

Total : $282\,\text{Cal}$

The percentage of calories from fat :

$\dfrac{126\,\text{Cal}}{282\,\text{Cal}} \times 100\% = 45\,\%$

c) Jerseymaid Peach Yogurt (lowfat)

Fat:

$$2.5\,g \times \frac{9\,Cal}{1\,g} = 23\,Cal$$

Protein:

$$9\,g \times \frac{4\,Cal}{1\,g} = 36\,Cal$$

Carbohydrate: $47\,g \times \frac{4\,Cal}{1\,g} = 188\,Cal$

Total: 247 Cal

The percentage of calories from fat:

$$\frac{23\,Cal}{247\,Cal} \times 100\% = 9\%$$

63. Answers will vary by person, a sample calculation is done below with a 130 lb female who does moderate exercise (approximately 300 Cal/hr) for 45 minutes a day, sleeps for 8 hrs a day and the rest of the time burns 90 Cal/hr.

$$130\,lb \times \frac{0.5\,Cal}{lb\,hr} \times 8\,hrs = 520\,Cal\ burned\ while\ sleeping$$

$$45\,min \times \frac{1\,hr}{60\,min} \times \frac{300\,Cal}{1\,hr} = 225\,Cal\ burned\ while\ exercising$$

$$15.25\,hours \times \frac{90\,Cal}{1\,hr} = 1373\,Cal\ burned\ the\ rest\ of\ the\ day$$

Total Calories burned in 1 day = 2118 Cal

65. Answers vary by person. (Sample calculation done with the data from problem 63)

$$\frac{3\,lb}{1\,month} \times \frac{1\,month}{30\,day} \times \frac{3500\,Cal}{1\,lb} = 350\,Cal\,/\,day$$

Because the person expends 2118 Cal / day, the intake that would provide a 350 Cal increase is

2118 + 350 = 2468 Cal

67. The amount of excesses calories burned per day is:

2200 Cal/day – 2000 Cal/day = 200 Cal/day

164

$$5.0 \text{ lbs} \times \frac{3500 \text{ Cal}}{1 \text{ lb}} \times \frac{1 \text{ day}}{200 \text{ Cal}} = 88 \text{ days or approximately 3 months}$$

69. The difference in weight is:

$$250 \text{ lb} - 170 \text{ lb} = 80 \text{ lb}$$

$$80 \text{ lb} \times \frac{0.5 \text{ Cal}}{\text{lb hr}} \times \frac{24 \text{ hr}}{1 \text{ day}} = 960 \text{ Cal / day}$$

71. $\dfrac{10 \, \mu g}{1 \text{ day}} \times \dfrac{1 \text{ egg}}{0.65 \, \mu g} = 15 \text{ eggs / day}$

SOLUTIONS TO FEATURE PROBLEMS:

75. Vitamins A and D are fat soluble and vitamin C is water soluble. Vitamin C is different from the others because is has multiple –OH groups on it which can hydrogen bond with water making vitamin C water soluble. Vitamins A and D do not have any –OH groups that can hydrogen bond with water and they both have large hydrocarbon chains on them making them fat soluble.

REVIEW TESTS FOR CHAPTER 18

Match the following statements or phrases to the end-of-chapter terms.

1. The class of carbohydrates that include the monosaccharides and disaccharides.

2. The body's primary fuel and the exclusive fuel of the brain.

3. A disaccharide consisting of the monosaccharides glucose and galactose.

4. A condition where adults with low lactase levels have trouble digesting milk or milk products.

5. The complex carbohydrates which are the structural material of plants and trees.

6. Proteins that contain all the essential amino acids in the right proportion.

7. Triglycerides which are a combination of glycerol with three fatty acids.

8. Specialized proteins that act as carriers for fat molecules and their derivatives in the bloodstream.

9. This classification of vitamins include A, D, E, and K.

10. The vitamin important in vision, immune defenses, and maintenance of body lining and skin.

11. This vitamin serves as an antioxidant in the body.

12. Vitamin C is involved in the synthesis of this type of connective tissue.

13. Vitamin C deficiency results in this condition.

14. The second most abundant mineral in the body and is largely bound with calcium in bones and teeth.

15. The main element involved in bodily fluid level regulation.

16. This minor mineral composes a critical part of hemoglobin.

17. The class of stabilizers added to foods to keep them moist.

18. The type of insecticides which are primarily broad-spectrum insecticides.

19. The kind of herbicide that works by causing plants to lose their leaves.

20. A 1:1 mixture of 2,4,5-T and 2,4-D herbicides.

ANSWERS TO MATCHING

1. simple carbohydrates
2. glucose
3. lactose
4. lactose intolerance
5. fiber
6. complete proteins
7. fats and oils
8. lipoproteins
9. fat-soluble vitamins
10. Vitamin D

11. Vitamin E
12. collagen
13. scurvy
14. phosphorus
15. sodium
16. iron
17. humectants
18. organophosphate insecticides
19. defoliants
20. Agent Orange

SELF-TEST QUESTIONS

Completion: Write the word, phrase, or number in the blank that will complete the statement or answer the question.

1.   Simple carbohydrates are easily and efficiently transported in the bloodstream to all areas of the body due to the presence of _____ groups on their structure.

2.   What is the name given for the chemical reaction shown below?

sucrose + water → glucose + fructose

3.   The body can metabolize proteins for energy, providing about _____ Cal/gram.

4.   The main carriers of blood cholesterol are _____ lipoproteins.

5.   Complete the following equation:
Energy intake = Energy _____ + Energy_____

6.   The deficiency of the trace mineral _____ results in goiter, a condition in which the thyroid gland swells.

7.   All food additives are monitored and regulated by the _____.

8.   Emulsifiers are added to food to keep mixtures of _____ and _____ substances together.

9. The chemical stability of chlorinated hydrocarbons allow them to accumulate in soil and water supplies, thus moving up the food chain, a process called _____.

10. A particular instant breakfast bar lists the following nutritional information:

Fat	5 g
Protein	3 g
Carbohydrates	22 g

What percentage of the total calories is obtained from fat?

Multiple choice: (Select the correct answer from the choices listed.)

11. Which one of the following statements concerning starch is <u>false</u>?

   a) Starch is composed of glucose units liked via beta-linkages.
   b) One source of starch is corn.
   c) Our bodies can digest starch.
   d) Starch stores energy that plants need for growth.

12. What is the biochemical classification for cholesterol?

   a) carbohydrate
   b) lipid
   c) steroid
   d) protein

13. What vitamin promotes the absorption of calcium through the intestinal wall and into the blood?

   a) A
   b) E
   c) C
   d) D

14. Which one of the following carbohydrates is classified a disaccharide?

   a) sucrose
   b) fructose
   c) galactose
   d) glucose

168

15. Sodium nitrite is classified as what type of food additive?

    a) antioxidant
    b) antimicrobial agent
    c) anticaking agent
    d) emulsifier

16. The primary nutrients that a plant gets from soil are potassium, nitrogen, and
_____.

    a) phosphorous
    b) magnesium
    c) calcium
    d) sodium

17. Consider the following structure of the herbicide, 2,4,5-T. The arrow points to what functional group?

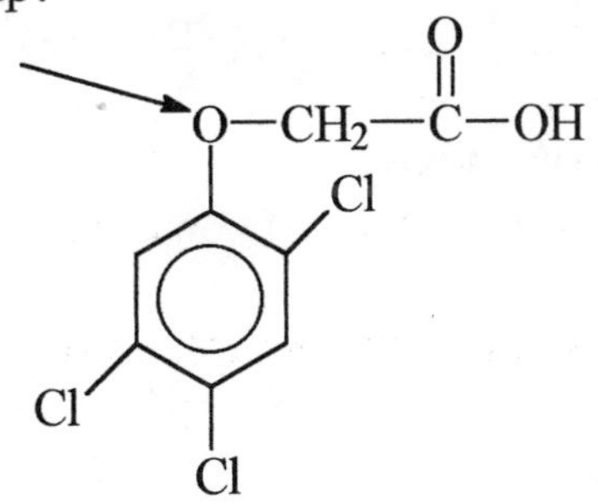

    a) ester
    b) ketone
    c) ether
    d) carboxylic acid

18. All of the following compounds represent fertilizers except

    a) $Ca(H_2PO_4)_2$
    b) $CaSO_4$
    c) NH_4NO_3
    d) $CO(NH_2)_2$

19.   What does basal metabolism represent?

a)   the minimum expenditure of energy utilized in all activities involved in vital functions.
b)   the minimum expenditure of energy utilized only in maintaining a constant body temperature.
c)   the minimum expenditure of energy utilized only when sleeping.
d)   the maximum expenditure of energy utilized in all activities involved in vital functions.

20.   Which of the following insecticides can accumulate in the environment and bioamplify in the food chain?

a)   carbaryl
b)   aldicarb
c)   malathion
d)   DDT

ANSWERS TO SELF-TEST QUESTIONS

1.   OH or hydroxy
2.   hydrolysis
3.   four
4.   low-density
5.   expended, stored
6.   iodine
7.   FDA or Food and Drug Administration
8.   polar, non-polar
9.   bioamplification
10.

Fat	5 g	×	9 Cal/g	=	45 Cal
Protein	3 g	×	4 Cal/g	=	12 Cal
Carbohydrates	22 g	×	4 Cal/g	=	88 Cal
Total					145 Cal

$$\frac{45 \ \cancel{Cal}}{145 \ \cancel{Cal}} \times 100\% = 31\%$$

11.   a
12.   c
13.   d

14.   a
15.   b
16.   a
17.   c
18.   b
19.   a
20.   d

CHAPTER 19

NANOTECHNOLOGY

ANSWERS TO QUESTIONS:

1. A nanometer is 10^{-9} meters.

3. Medicine is one of many fields that would benefit from extreme miniaturization. There would be many advantages if machines could be made smaller to navigate in the blood, unclog blocked arteries, destroy cancerous cells and even perform surgeries inside the body. The possibility for miniature electronic devices is also appealing since you could have all the luxuries of the most powerful computers in the palm of your hand.

5. Nanotechnology has looked possible since the discovery of the scanning tunneling microscope that can image and move individual atoms. This and other recent advances such as the discovery of buckyballs and buckytubes have helped move nanotechnology from merely an idea into the laboratory.

7. The scanning tunneling microscope was discovered in 1981 by Gerd Bennig and Heinrich Rohrer at the IBM Zurich Research Laboratory.

9. Tapping or oscillating while taking an image the AFM further increases its resolution and minimizes damage to samples. By tapping the AFM cannot only image metallic surfaces but also biological samples.

11. Buckminsterfullerenes are C_{60} and have the same structure as a soccer ball.

13. Nanotubes have already been used to act like a tiny scale weighing objects as small as viruses, they have acted as atomic pencils, and as wire, connecting two parallel electrodes. Nanotubes also have the possibility of being used to desalinate water and make flat-panel displays.

15. An artificial red blood cell would be a nanovesicle pumped full of oxygen. The nanovesicle, at the first sign of the cardiovascular system failing, would begin the slow release of oxygen to the body and allow it to survive until the cardiovascular system is restored.

17. Nanotechnology although it looks promising has obstacles to overcome before it becomes a reality. Nanomachines cannot be built in the same way as full-sized machines and have things like intermolecular forces to contend with. Engineers of nanomachines have yet to build one so skepticism lingers and will probably stay long after one is constructed in order to discuss the ethics of it all.

SOLUTIONS TO PROBLEMS:

19. $7.8 \times 10^{-6} \, m \times \dfrac{1 \, nm}{1 \times 10^{-9} \, m} = 7.8 \times 10^3 \, nm$ or 7800 nm

21. $0.10 \, mm \times \dfrac{1 \times 10^{-3} \, m}{1 \, mm} \times \dfrac{1 \, nm}{1 \times 10^{-9} \, m} = 100{,}000 \, nm$

$\dfrac{100{,}000 \, nm}{10 \, nm} = 10{,}000$ nanotubes will fit in the width of a line

23. hexagons

REVIEW TESTS FOR CHAPTER 19

Match the following statements or phrases on the left to the term on the right.

1. The science of making things extremely small.

2. A microscope capable of atomic resolution on metallic surfaces.

3. A microscope capable of atomic resolution on non-metallic surfaces.

4. A new form of carbon discovered by Richard Smalley.

5. Long thin tubes of carbon atoms.

nanotube
buckyball
scanning tunneling microscope
nanotechnology
atomic force microscope

ANSWERS TO MATCHING

1. nanotechnology
2. scanning tunneling microscope
3. atomic force microscope
4. bukyball
5. nanotube

SELF-TEST QUESTIONS

Completion: Write the word, phrase, or number in the blank that will complete the statement or answer the question.

1.   A scanning tunneling microscope works by measuring the _____ _____ between an atomically fine tip and a metal surface.

2.   An atomic force microscope works by measuring the deflection in a _____ _____ caused by atoms or molecules deflecting the cantilever tip.

3.   Buckyballs contain ____ carbon atoms bonded together to form a hollow sphere.

4.   Nanotubes can be used as _____ in microscopic circuits.

5.   In medicine, Nanotechnology may one day produce artificial _____.

Multiple Choice: Select the correct answer from the choices listed.

6.   Extreme miniaturization is at least possible because it already exists in:

   a)   computer technology
   b)   living organinisms
   c)   transistors
   d)   automotive technology

7.   The scientist(s) that discovered the scanning tunneling microscope:

   a)   Bennig and Rohrer
   b)   Smalley, Curl, and Kroto
   c)   Watson and Crick
   d)   Fleming

8.   Artificial cells that can carry oxygen can keep a person alive in the event of:

   a) liver failure
   b) kidney failure
   c) heart failure
   d) colon failure

9.   The scanning tunneling microscope can produce images of:

a) protons
b) neutrons
c) electrons
d) atoms

10.   To image samples that are prone to damage, the tip of an atomic force microscope is:

a) lifted from the surface
b) coated with plastic
c) tapped on the surface
d) not used

ANSWERS TO SELF-TEST QUESTIONS

1.   electrical current
2.   laser beam
3.   60
4.   wires
5.   cells
6.   b
7.   a
8.   c
9.   d
10.   c